MW00621005

Night
Magic

ALSO BY LEIGH ANN HENION

*Phenomenal: A Hesitant Adventurer's Search
for Wonder in the Natural World*

Night
Magic

*Adventures Among
Glowworms, Moon Gardens,
and Other Marvels of the Dark*

LEIGH ANN HENION

Algonquin Books of Chapel Hill 2024

Published by
ALGONQUIN BOOKS OF CHAPEL HILL
an imprint of Workman Publishing
a division of Hachette Book Group, Inc.
1290 Avenue of the Americas
New York, NY 10104

Portions of "Fireflies Blinking" appeared in different form in the *Washington Post*.
Permission to reprint "To Know the Dark" granted by Wendell Berry.

Printed in the United States of America.
Design by Steve Godwin.

Library of Congress Cataloging-in-Publication Data
Names: Henion, Leigh Ann, author.
Title: Night magic : adventures among glowworms, moon gardens, and other marvels
    of the dark / Leigh Ann Henion.
Description: First edition. | Chapel Hill : Algonquin Books of Chapel Hill, [2024] |
    Includes bibliographical references | Summary: "In a glorious celebration of the
    dark, nature writer Leigh Ann Henion invites us to leave our well-lit homes and
    step outside to embrace the biodiversity that surrounds us"—Provided by publisher.
Identifiers: LCCN 2024021972 (print) | LCCN 2024021973 (ebook) |
    ISBN 9781643753362 (hardcover) | ISBN 9781643756196 (ebook)
Subjects: LCSH: Henion, Leigh Ann. | Night gardens. | Night. | Phenogodidae. |
    Fireflies.
Classification: LCC SB433.6 H46 2024 (print) | LCC SB433.6 (ebook) |
    DDC 635.9/53—dc23/eng/20240617
LC record available at https://lccn.loc.gov/2024021972
LC ebook record available at https://lccn.loc.gov/2024021973

10 9 8 7 6 5 4 3 2 1
First Edition

For Archer, my adventure partner,
    and for everyone
    who has ever felt lost in the dark

How insupportable would be the days,
if the night with its dews and darkness
did not come to restore the drooping world.

—HENRY DAVID THOREAU

To go in the dark with a light is to know the light.
To know the dark, go dark. Go without sight,
and find that the dark, too, blooms and sings,
and is traveled by dark feet and dark wings.

—WENDELL BERRY

Contents

Night
Magic

Preface

~~~~~~

Years ago, I heard a story about a boy who was lost, alone, in the woods. When a man from the search-and-rescue team found that child, curled like a baby rabbit among leaf litter, he did not lift him from the forest floor. He did not whisk the boy away to an artificially lit, human-built space. Instead, he lowered himself to the ground beside the boy and asked him to describe what he had heard and seen and felt during the long hours he'd spent in the dark, terrified.

At his rescuer's prompting, the boy described the sounds of creatures cackling, the feel of tiny feet tickling his skin. In return, the man, to the best of his ability, explained what had been going on around him. That first responder understood that, for the rest of the boy's life, his memory would wander back to that place. And, if he had been properly introduced to the creatures he'd encountered there in the dark, he would not have to continue living in fear of them.

Unfortunately, most of us are rarely advised to hold space for darkness, and it's difficult to find guides that might help us better relate to it. Against darkness, the Western world has a deep cultural bias. Almost every storyline we're familiar with suggests that we should banish it as quickly as possible—because darkness is often presented as a void of doom rather than a force of nature that nourishes lives, including our own.

But darkness is an integral and essential part of the human experience, and it's one that we are collectively losing. Organizations ranging from DarkSky International to the American Medical Association have implored the public to fight light pollution, which has been shown to cause increased rates of diabetes, cancer, and a variety of other ills, as well as degradation to entire ecosystems. Still, light pollution continues to grow.

What might we discover if we pause to consider what darkness offers? What might happen if we, as a species, stopped battling darkness—negatively pummeled in popular culture and even the nuance of language—as something to be conquered and, instead, started working with it, in partnership?

This is the story of how I set out to re-center darkness by spending time with some of the diverse and awe-inspiring lifeforms that are nurtured by it. We are surrounded by animals who rise with the moon, gigantic moths and nocturnal blooms that reveal themselves incrementally as light fades. In the past, I might have conceptualized a journey to align with natural darkness as requiring a jaunt to the Canary Islands, the darkest place on Earth. I might have neglected the nocturnal expanse of my Southern Appalachian homeland. But we are increasingly in need of models of how to find wonder on our own patch of planet. In this way, I hope my quest will strike curiosity that can be applied by anyone, to any nocturnal landscape.

Darkness turns familiar landscapes strange, evoking awe by

its very nature, in ways that meet people wherever they stand. In Appalachia, as everywhere, night offers a chance to explore a parallel universe that we can readily access, to varying degrees. Nocturnal beauty can be found not only by stargazing into the distant cosmos or diving into the depths of oceans, but by exploring everyday realms of the planet we inhabit.

We, along with everyone we know, have relationships to darkness that influence how we think about it, talk about it, and move through it. But, unlike that child found in the woods, we're rarely given opportunities to contemplate our experiences. Whether we are swimming in bioluminescent tides off the coast of California or watching iridescent moths hover over a sidewalk in Brooklyn, nearly anywhere on Earth can—at the flip of a switch—become a wilderness of possibility.

As you travel with me through the fern-sprouting valleys and cloud-cresting peaks of Appalachia, encountering creatures both familiar and strange, I hope you will join me in recognizing darkness as a restorative balm for this burning world. And by the time you've turned the last page, I hope you won't feel the impulse to always quicken your step when encountering darkness, imagining perils. Instead, when you come across shadows, I hope you'll be inspired to sometimes slow your stride, alert to marvels.

Season of
Inspiration

Fireflies Blinking

~~~~~~

Synchronicity

I've been in Great Smoky Mountains National Park for less than an hour when I'm mistaken for a woodland fairy. Even though I'm here to witness the ethereal phenomenon of synchronous fireflies—a species famed for its ability to flash in unison—the association is surprising since I'm feeling more like a haggard dweller of the modern world than an enchanted being of old-world mythology. In fact, when I hear a stranger calling out from across the forest glen I'm wandering, it takes me a second to realize that she's addressing me. She waves me over and asks again: "Are you a magical creature?"

The woman gestures toward the two young children with her and says, "We saw you walk down to the river, and then you disappeared. I told the girls you must be magical. This whole place is magical. Reminds me of Narnia or something."

It does feel like we've traveled through a portal to another realm. The woman is sitting on a porch stoop, but there's no porch. And there's a chimney nearby, but no house. To reach the

Elkmont-area trailhead, we—along with hundreds of other visitors here to witness the synchronous fireflies' light show, which generally occurs in a two-week period around early June—had to walk through an avenue of mountain cabins, abandoned after the park was formed. Remnants of the former human settlement—some of which has been lost to the elements—are visible everywhere, scattered among river-rounded stones and beds of fern.

In 2021, Tufts University released the first-ever comprehensive study of firefly tourism. They found that, globally, one million people travel to witness firefly-related phenomena every year. Given that the synchronous fireflies of Elkmont are some of the most famous fireflies in the world—and that I live in Southern Appalachia, their home region—coming across the study during a pandemic lockdown made me think it was past time for me to see these brilliant creatures.

The firefly event in Great Smoky Mountains National Park, which straddles this section of Tennessee and North Carolina, draws visitors from across the continent. Years ago, the National Park Service instituted a lottery for people to secure passes since the species' growing popularity raised concerns about conservation. I explain to the woman that I'd dipped down to the river for a brief respite from the crowd. She empathizes. Even with attendance limitations, the annual gathering isn't a small one.

I'm here hoping to glimpse fireflies' bioluminescence, or living light, partly because I've been spending too much time basking in the illumination of screens. I've fallen under the influence of phones, computers, and tablets. For several seasons now, I've been beating myself against screens like a moth against a lightbulb, seeking entertainment that might numb me, news that might serve to comfort me. In a time of global confusion, I've been trying to find answers that do not exist. The process has only served to disrupt my animal instincts—and the influence of artificial light in my life isn't limited to electronics.

According to DarkSky International, 99 percent of people in the United States live under the influence of skyglow—diffuse, artificial brightening of the night sky—with a loss of unfettered access to the blinking sun-and-moon patterns with which we evolved. Internationally, light pollution is increasing at exponential rates with no signs of slowing. It's as if we, as a species, have grown afraid of the dark. Tonight, I'm hoping to break the spell that artificial lights have cast.

~~~~~~~~~~

Along the trail designated for firefly viewing, people have been setting up folding chairs as if they're waiting for a parade. They're a diverse bunch. Despite the awkwardness of talking to strangers in the dark, I learn that there are nine-month-olds and ninety-year-olds among them. Some of them have been to the firefly viewing several times. Some, from the West Coast, are awaiting the first firefly sighting of their lives. They've come alone. They've come with their families. They've come because this event is something they've always wanted to attend and, due to the state of the world, they've stopped taking next year for granted.

Firefly habitat is so specific, so mercurial, that it's possible to see a great show from one section of the trail while another remains relatively dark. No one, not even rangers, can predict the best seats for the evening, so people mill around until they find a spot that feels right to them. Finally, dusk comes.

When the first synchronous fireflies appear, sporadically flashing, they don't seem, to my untrained eye, to be much different from common species that illuminate backyards across the country. But as their numbers grow, expectant murmurs travel up and down the row of spectators. Instinctually, when hundreds of insects grow to be thousands—each appearing to light the next in line, like a candle being passed—the crowd stands.

For a while, the insects' rhythms remain a bit discordant, like those of an orchestra warming up. Scientists have found that the more individuals participate, the more in tune the insects become. Before long, it's clear that the fireflies are working in unison. The effect isn't a lights-on-lights-off situation, as I'd expected; it's more like watching one of those raised-hand stadium waves, when people at a sporting event sequentially lift their hands, swept into the fervor of something larger than themselves.

The insects are responding to each other's light, working with their neighbors to find their role in the whole. From a distance, the activity appears as a shimmering current of light running through the forest from right to left: Whoosh. Then darkness. Then again, a whoosh of light.

I cannot see the face of the woman beside me, but I come to attention when she calls out, "*Dun, dun, dun, dunnn,*" mimicking Beethoven's famous symphony motif. "It's like they're playing music," she says to someone beside her.

Spontaneously, I pipe up: "I couldn't help overhearing what you just said about music. Have you heard about how the synchronous fireflies were found here?"

"Whoa, a messenger from the dark!" she says, laughing. "No, tell us!"

So I share what I've heard, about how naturalist Lynn Faust, who used to spend summers in the now-defunct Elkmont community, grew up admiring the fireflies we're watching. As an adult, she came across an article about synchronous-flashing fireflies in Asia, and she recognized similarities in what the scientists were reporting and what she'd seen as a child.

When she reached out to researchers in the 1990s, they were skeptical that an unknown-to-science species existed in the most-visited national park in the country, so she sent a musical composition mimicking the sequence of flashes in Elkmont. It's what convinced firefly scientists that they should make the trip to Great Smoky Mountains National Park, where they confirmed a never-before-recorded synchronous species: *Photinus carolinus*. This is, ultimately, how we all ended up here, bearing witness this evening.

I can sense more people gathering around me as I'm speaking. When I finish, strangers' voices ping to my left, to my right, from the trail behind me. Their words ring like bells.

"Amazing!" says a baritone.

"Fantastic!" shouts a soprano.

"What, exactly, do you think they're singing?" a man asks the crowd.

"Beyoncé! 'All the Single Ladies'!" a woman says. Laughter ripples up and down the trail.

Most people in attendance seem to be familiar with the concept of firefly flashes as a function of courtship. The insects we're seeing are males, signaling to females who stay close to the ground. Scientists generally agree about the utility of fireflies' bioluminescence as mating-related, but they've long tussled over how, exactly, fireflies make light. It's generally thought that illumination occurs when fireflies send nerve signals to their lanterns—allowing oxygen to ignite inborn organic compounds in their bodies. This means, in a roundabout way, that when you see fireflies light up, you're watching them inhale. And these fireflies are pulsating like cells of a glowing, forest-size lung.

Collectively, the crowd gasps and sighs as fireflies crackle. But despite the dazzle, I find my eyes drifting downward toward the infinitely dark ground. Because once I started researching fireflies, I came across this unshakable fact: By the time we see fireflies in flight, they have potentially been living among us for up to two years in various life stages, dimly glowing on the ground. What we're witnessing now is the grand finale of a long-term metamorphosis. These famed fireflies have spent much of the past year crawling around in the dark to find what they needed to survive so that their species might ultimately thrive.

~~~~~~~~~

There are more than 2,500 known species of fireflies in the world, and 19 of those—with synchronous being the most famous—reside within the borders of Great Smoky Mountains National Park. Will Kuhn, Director of Science and Research at Discover Life in America, a nonprofit centered on biodiversity, believes there are even more. "I don't think we've found all firefly species in the park," he says. "And there's still a lot we don't know about the ones we have found." If that's true, there is a chance we won't know what we've got even after it's gone. Globally, firefly populations are under assault, and the largest threats to their well-being—according to the Tufts report—are habitat loss, pesticide use, and light pollution.

When I meet Will, the day after the Elkmont event, he's chatting with two dozen people who've signed up for a synchronous firefly viewing party hosted by his organization, which often partners with universities and other research institutions. Since Discover Life in America's founding in 1998, the group's efforts have led to the documentation of more than 10,000 animal and plant species in Great Smoky Mountains National Park—with more than 1,000 of those being previously unknown to science.

We're getting ready to travel down the mountain to a private creek-side habitat outside of the national park. Will knows Norton Creek to be home to a large population of synchronous fireflies, which have now been observed in Appalachia as far north as Pennsylvania. The group is already buzzing with questions. Many of them have turned to Discover Life in America because, year after year, they've failed to win federal lottery passes. Since they've found another route to witness the synchronous phenomenon, they're already feeling lucky.

One of the women encircling Will says that she's excited for a good show, because she only has "plain old fireflies" on her farm in Ohio. Will suggests that if she does a little research, she'll find that her region is likely home to several species, each with their own songs. The most common firefly in the United States is the big dipper, but there are 150 species across the country, all with their own specific habitats and behaviors. Each firefly species that flashes has patterns that are as unique as fingerprints. And where you find one species in a meadow, there's a good chance you'll be able to find others in forests nearby. Diverse habitats breed diverse kinds of light.

Susan George, a nurse from San Antonio, Texas, lives in the city proper, and she's always been amazed that fireflies are tenacious enough to find homes there, in rare squares of land that have been spared from asphalt and concrete. "Sometimes when I'm sitting out in my yard, fireflies land right on me," she says.

The farmer from Ohio nods. "When they do that," she says, "it feels like love."

Susan gives a weak smile. "I'm here because, at the hospital, I work with bugs of a different kind," she says. "And I really needed a break." Everyone falls silent. We are—as any group of humans might be in the wake of a pandemic—a swarm of loss embodied.

Unfortunately, when we finally make it to the waterway, the local population of synchronous fireflies fails to greet us. There are only a few partnered dots of light. Predicting emergence dates of fireflies at Norton Creek involves, as it does everywhere, a formula of temperature patterns and other factors. But even with careful calculation, the details of firefly metamorphosis can be difficult to precisely predict. It's several degrees cooler here than it has been in the Elkmont region of Great Smoky Mountains National Park. The synchronous residents of Norton Creek apparently need a few more days to fully wake.

Long after it's clear that we've been stood up, the group continues to loiter at the edge of the woods. Just when it seems spirits are irretrievably waning, someone spots a strange orb of light rising from the understory. It peers at us from across the creek, blue and unblinking.

I've been familiar with the term "blue ghost" for years. Until recently, though, I didn't understand that synchronous and blue ghost fireflies were different species. They have slightly different mating seasons, but these often overlap as conditions transition from evening to evening. Currently, on Norton Creek, it's the blue ghost population that's peaking.

The ghost moves toward us, floating more than flying.

Soon, there are carpets of light in and around the forest

on all sides of us. These creatures, notable for their neon-blue color and enduring flashes—which hold for up to sixty seconds at a time—are visible demonstrations of how to breathe deeply. Their traceable flight patterns make them look as though they're intoxicated.

As people scatter, I find myself walking alone. But with every step I take, more fireflies reveal themselves, until the entire mountain is trembling. Blue orb-fairies, hundreds of them, appear to be following me. They're continuously swooping and swerving and serenading me—not as a visitor to this landscape, but as part of it.

I've seen the aurora borealis in the Arctic. I've witnessed migrations in the Serengeti. I've snorkeled the Great Barrier Reef in Australia. Yet I'm not sure that I've ever appreciated any natural phenomenon more than this marvel of Appalachia.

By the time I hear voices on the road ahead, I've lost all sense of time and space. In dim moonlight, I can make out half a dozen silhouettes in the distance. Will's voice is hushed. "People typically don't walk around at night without lights on," he says. "But it's amazing what happens when you let your eyes adjust to the dark, when you take the time to really look."

Rising

For weeks after I return home, I find myself scanning meadows and creeks—not as scenic backdrops, but as habitat. Every plot of land I see is suddenly weighted with potential glory. And each night, around 9:30 p.m., when I'd typically be logging on to

Netflix, I get the urge to go outside to check on the local firefly population.

The mountains of North Carolina, where I live, are primed to become the next firefly tourism hot spot. In 2019, a population of synchronous fireflies was discovered on Grandfather Mountain, a beloved regional attraction near my hometown of Boone. The entomologist who found the species—during a nocturnal stroll, taken on a whim—had been traveling a trail I've walked dozens of times in daylight.

The mountain is closed to night visitors, but natural resource staff members are investigating how Grandfather might host future firefly viewings without harming habitat. Surveys of synchronous populations on the mountain have led participants to find—in locales frequented by more than half a million visitors a year—previously overlooked blue ghost populations as well.

Even my ordinary front door, on the far end of neighboring Watauga County, is a portal to a parallel universe once the sun has set, though I hadn't previously recognized it. My guided night walks in Great Smoky Mountains National Park have acted as training. Even so, it still takes me a few nights of distant firefly watching to leave the familiarity of my front porch. This ease-in approach gives me an opportunity to correct the light pollution seeping from my house, mitigating trespasses against my bioluminescent neighbors that I hadn't been aware I was making. I close curtains, turn off porch lights. The difference made by these small changes is staggering.

As the darkness around my house deepens, I move farther out. I take to sitting beyond an old chicken coop, watching what I now understand to be femme fatale fireflies winking from tree-tops and big dippers plunging through meadows. Then, one night, I decide that I'm going to leave my immediate environs to explore the valley beyond.

I set out for a place where fields and forest meet. When I reach a neighbor's livestock gate—open since its last inhabitants, a family of goats, were killed by an unidentified predator—I pause, mustering the courage to enter the rhododendron thicket in front of me, beyond briars where I often see rabbits munching and jumping. But before I embark on my chosen path, I hear rustling in the feral pasture above me.

My eyes are not fully attuned, but they're adjusting. I use vestiges of twilight to trace the ragged outline of high grass on the hill. I'm on the verge of dismissing the sounds as manifestations of anxiety when a bobcat streaks through the air. I can see the animal arched like a crescent moon that rises and sets, nearly close enough to touch.

The predator has pounced onto something I cannot see—so quickly that I hardly have time to register what's going on. Then, from thorny bramble, the bobcat exhales in a guttural hiss. The sound slithers around me, and I yelp from the pressure of it.

I turn to run, but somewhere beyond my conscious mind, I have a vague understanding that fleeing would trigger the animal's prey instinct. It takes everything I have to resist running. I pivot to an alternate route, keeping my pace steady, targeting the pool of a distant security light, even though I know the light cannot save me. When I realize this, I mutter to myself: *The light cannot save you.* That's when it registers: I might have set out on my firefly pilgrimage because I wanted to revel in light, but what I needed was a reconciliation with darkness.

Fireflies are light bearers, but—blue ghosts notwithstanding—it is the darkness between most species' flashes that reveals their true character. Without intermittent darkness, there would be no firefly music, no signal, no communication. There would be no synchronized light shows, no J-stroke patterns from the common big dipper. There would only be glare. Stars are, after all, in the sky above us, even at midday, but we only see them when the sun takes its leave. Because, while it's true that only light can drive out darkness, there are some forms of light that only darkness can reveal.

We live in an age that's asking us to get comfortable with constant disruption. There will always be, as there always have been, threats beyond our line of sight. But as we venture into the unknown, we also stand to encounter wonders yet unimaginable. Of this, I needed to be reminded. I keep walking.

When I'm half a mile from the site of my bobcat encounter, I slow my pace. Out of the corner of my eye, a lone firefly is blinking in what appears to be the synchronous species' recognizable pattern. It repeats, with a dark pause that holds, beat after beat. I cannot imagine that I've found a synchronous firefly, but I'm no longer willing to discount the potential of any patch of land in Southern Appalachia.

There are, in my best estimation, half a dozen firefly dialects being spoken here. These creatures are getting their bearings in a place that, up until now, they've only seen from ground level. We are simultaneously gaining new perspective of our shared habitat. In their presence, my shoulders—long tense and recently tightened by the breath of a bobcat—begin to soften. I realize that, in facing the fight-or-flight unknowns of night, even my hand muscles have clenched. I consciously unfurl my fists to accept darkness as a gift.

It was only after I started communing with fireflies that I discovered that foxfire, the region's famed bioluminescent fungi, isn't a single species, as I've always thought; it's a term used for a myriad of glowing organisms that people once collected from the forest floor to serve as lanterns. These mountains are full of hooting owls and entire landscapes that bloom only when touched by the moon. I'm beginning to know darkness, for the first time in my life, not as a realm of unknowns best avoided, but rather as a potential state of enchantment. I begin to wonder: If fireflies can be unexpected mentors of meaning, what might moths and salamanders and the other living marvels of this dark landscape have to teach?

Slowly, entire constellations of fireflies rise from the coal-black earth around me, twinkling with oxygen. I attempt to align with their rhythm: Inhale, light. Exhale, dark. We are breathing, in sync, on this complicated planet. And even the deepest parts of the valley I'm standing in are pulsating with life, illuminated.

Spring

Salamanders Migrating

Under a Rock

It's been at least a week since my backyard saw a blizzard, though snowbanks are only now beginning to soften. The largest is a four-foot monument that's transforming into meltwater. It has carved a small channel in my unpaved driveway. The steady stream is flowing past an old apple tree, toward the New River at the bottom of my yard, with currents that will carry it through mountains toward the ocean. It's delightful to watch those hard-packed snowflakes find release. But for me, the space between winter and spring still feels a little like purgatory.

I have a vague notion that this is when the underworld of soil begins to stir with spotted salamanders. They have, like me, been waiting for warmer seasons. These creatures spend their entire lives underground, outside of a few nights each spring when, across North America, they emerge in search of ephemeral, or vernal, pools, which are fed by snowmelt and rain. These pools disappear and reappear annually, giving them a place to breed away from hungry fish. Salamanders are animals who depend on impermanence.

To my human senses, the promises of spring are mostly abstractions. Seasonal action is happening in dark places I cannot access. But finding ephemeral pools might be a way to catch a glimpse of the spring-forward motion I'm craving.

I am ready to welcome longer hours of daylight, but I've been considering the way salamanders will mark the coming season without directly feeling sun on their skin. They live in darkness and travel to those seasonal pools only thanks to nocturnal slipstreams.

~~~~~~~~~~

The mountains have already begun to nudge animals out of their burrows, soil freezing and contracting in push patterns. I know that some of my neighbors go out each spring to see salamanders migrating toward ephemeral breeding pools, and I start to wonder how I might join them. I begin to muse about how better understanding the way salamanders relate to night might help me more nimbly understand my own relationship to it—as well as the enduring human impulse to banish darkness before it can take root.

To suggest that someone is "living under a rock" or that they're "in the dark" means that they aren't aware of what's going on around them. Yet most of us have little knowledge of what lives in the dark, under actual rocks—maybe because sayings like this indicate that they aren't worth much. But turn over a stone in any Appalachian woodland and you'll find all sorts of writhing life-forms.

Salamanders are spotted and striped, so lovely that it isn't uncommon for people to refer to them as jewels. Of 550 known species in the world, 77 can be found in Appalachia, and much of North America is rich with them. These creatures might spend most of their lives under rocks, but they are acutely aware of things we generally ignore, and their bodies bear wisdom that science is only beginning to investigate.

Finding a salamander in a creek indicates that the water is clean, because salamanders, which sometimes breathe through their skin, are highly sensitive to pollutants. They're well-known in scientific circles for their ability to regrow tissue, organs, and limbs. James Monaghan, of Northeastern University, has been studying salamanders' regenerative abilities, which employ mechanisms that humans amazingly have in early developmental stages, though not in mature tissue. He suggests that human regeneration—guided by salamander clues—will realistically occur within the next decade.

Salamanders are masters of shape-shifting, but in the modern world, species that have survived 200 million years and three mass extinctions face a myriad of threats—among them, car tires. In 2015, Virginia's Center for Urban Habitats found that road-crossing salamander casualties can reach 50 percent. This has led to the formation of large-scale human "salamander brigades" in Iowa, New York, New Hampshire, and a multitude of other states, where people ferry animals to safety.

I don't know how to locate an ephemeral pool, but I soon hear from a biologist in the know that local salamander enthusiasts are already watching temperatures and gauging the rain that motivates salamanders to move. "We just have to be on call this time of year," she tells me, sounding every bit like a midwife waiting to hear from a woman going into labor.

Weeks into my attempts to figure out how I might witness this fleeting migration, a friend of a friend, Mike Gangloff, a biology professor at nearby Appalachian State University, reports that spotted salamanders have started to stir at a populous site. He's planning to take some of his students out, so he asks if I'd like to join them. It's just a casual patrol, meant to help salamanders cross from overwintering woodlands to their ancestral breeding pools. The site, as it turns out, is only a few miles from my house.

~~~~~~~~

Mike's nocturnal-wildlife-seeking adventures have taken him to some unexpected places—including the ditches behind a familiar fast-food restaurant, where he says salamanders are shockingly prevalent. But none of them seem more peculiar than our community's gun range, where he has, in the past, surveyed salamanders under a cacophony of gunfire. Tonight, the place is thankfully quiet.

As soon as he pulls up to meet me and the group of biology students who've convened, Mike jumps out of his truck to survey the roadbed, scanning at close range. "Wouldn't want to hit an animal!" he says.

Here, tires are weapons and guns have become unintentional tools of conservation. The daily booms of the firing range have deterred vacation-home owners and most locals from building houses and installing septic systems in the area. Because of this, there is little artificial light and an intact forest full of salamander burrows.

Mike leads us through a wetland that looks like a miniature diorama of the Appalachian chain, bumps and valleys gathering rain. Salamanders appear as ribbons of black weaving through sedge grass—and these are just the early party guests. "It's better the later it gets," Mike says. "It's like going to a bar. You want to be there late, when it starts rocking!"

The ground itself is communicating. Every time I lift my shoe it feels like I'm being released from a suction cup. *Thwap* goes my left foot as the boggy ground tries to hold it. *Thwap* goes my right.

One of the college kids is wearing open-toed shoes. He exhales sharply with each infusion of water into his sandal. Still, he doesn't complain. *Thwap*. "That," he says, "is a wonderful sound."

The location we're headed to, Mike explains, hosts a whole ephemeral pool complex. He suspects parts of it were built by beavers. "I've seen people sink right into the ground out here," he warns. "One minute they're there and the next minute they're not." It isn't quicksand, but it would be easy to lose a rain boot.

There are several shallow ponds and a deeper one, where I can see stray spotted salamanders floating, their dark bodies speckled with yellow-star dots. On the pond's silty bottom, there's a string of what look to be freshwater pearls. Mike points them out as spermatophores, packets of reproductive material that the males have left out for females to collect.

The females, who typically leave their mountain burrows a bit later than males, will ultimately take them into their bodies via a vent. Within weeks, these animals will be back underground, going about their business. "We don't know much about what salamanders do for the rest of the year," Mike says.

"They're just so chill," a student says when a male salamander comes into view, steadily marching toward the ephemeral pools. "Watch that little guy walk!"

The salamander does seem chill. His feet, which look like small purple hands, touch the earth two by two. *Left, right, left. Left, right, left,* every press and pull of his feet turning life's wheel.

Mike spots action below the surface of an ephemeral pool. "The females are starting to show up," he says. "It rained hard today. That's always a good sign of movement." Now that we have the lay of the land, we make our way from a grassy area to blacktop. Mike's colleagues have previously surveyed 400 spotted salamanders in a single night. But tonight the flow of salamanders is more of a trickle. Even so, it takes less than a minute for a student to notice an animal on asphalt.

If there were more salamanders to be handled, the student would likely be wearing gloves to prevent her from exposing the animals to human skin. But this is a low-key walk in the scheme of salamander crossings. Still, when touching a salamander, even to save it from other dangers, it's best to have wet hands.

The student, who has forgotten to bring a water bottle, presses her palm against the rain-soaked ground, absorbing what she can from the path. She inspects the glaze. Satisfied, she proceeds carefully. "The road is not for you, little one," she says, as she carries the animal to safety.

I keep pace with a teenager who is wearing pajama pants, as though he was unexpectedly beamed straight here from his couch. As it turns out, he isn't even in Mike's class. He was sitting around watching movies when he spontaneously decided to join his friends at the last minute. "They knew I'd be into this. My room's full of animals—lizards and plants, lots of them. People have started calling it 'the Sanctuary.'"

In our college town full of cinder-block dorms and

windowless apartments, his description sounds like an oasis for nature-deprived humans. It takes me a minute to realize that he's referring to his room as an animal sanctuary. "Salamanders are awesome," he tells me.

In recent years, scientists have discovered that woodland salamanders play a significant role in sequestering carbon. They are, according to biologists who work with them, unsung heroes in mitigating global warming. Their lifestyles lead to more carbon being stored in soil, with a significantly protective effect. In a rainy season, 170 pounds of carbon can be sequestered by salamanders in a single acre.

As a group, salamanders make up more living weight than any other creature in these woodlands. More than deer. More than Appalachia's famed black bears. "I think there are salamanders all around us right now," Mike says, studying a ditch. "We just don't see them, because they're masters of camouflage."

When one of the more squeamish students sees a salamander, she puts her cell phone on the road like a directional guardrail so she won't have to touch the animal. Her finger accidentally brushes the screen, and it lights up. Immediately, the salamander freezes.

From the salamander's perspective, this must be akin to being in Times Square with a larger-than-life electronic billboard. This creature has likely never been exposed to sunlight, much less artificial light at close range. The phone-holding student is still nervous to touch the animal, so I reach out to move the salamander away from the roadbed, but not before the group turns into paparazzi, their cell phones popping and flashing.

When my fingers close around the salamander's body, it almost feels like the animal can sense my unease in being a voyeur of the highest degree. But when her inky eyes meet mine, I cannot help but feel like she is smiling. I know it's just her resting amphibian face, but I can't get over the curve of her mouth, a grin that seems to permeate her entire body. Ultimately, I lay her in a ditch, away from car tires and selfie-seekers.

More salamanders are crossing the road in slow motion. *Left, right, left.* Half a dozen students watch for oncoming cars. But there isn't much traffic on this backroad. I start to wonder if our work is even helping. Then a car passes. In its wake, we find the pale outline of frog who didn't make it. "There's nothing to do for him," Mike says. Still, the students stand around the frog's body until one of them whispers, "This is sort of traumatic."

The mood shifts when we encounter another salamander just a few feet away. The student who had previously used her phone as a guardrail has, in the wake of our frog discovery, lost her inhibition. The stakes are now tangibly high. She reaches out to move the animal as the student in pajama pants coos, "These guys are too cute to be out here walking around on the *ground*!"

To chase salamanders is to endlessly contemplate the ground itself. The language of land—like the language of darkness— makes us strive for places aloft and alight, disparaging those down and dark. But the truth is, the darker soil appears, the more nutrients it generally holds, the more life it can support. Still, we are conditioned—almost all of us who have come of age going to zoos and aquariums—to think of the ground as somehow gross. We tend to value light over dark, high above low, the heavens as greater than the ground, perceived as entry to an underworld that many of my fundamentalist neighbors associate with hell.

A car approaches from a main road, and a passenger wearing

a headlamp jumps out to check the terrain before rolling forward. Our shift is over. This marks a transfer of the guard.

The animal lover in pajama pants pipes up to thank Mike for letting him tag along. This student arrived at the gun range seeking something beyond academic credit, something he still isn't sure how to articulate. Here, his only task was to hold the human world temporarily at bay so that salamanders could find their own way in a world that, unlike with his aquariums, he didn't have to create.

"I can't believe I never knew this was going on," he says, placing a hand over his heart in gratitude. "Walking around out here in the dark with salamanders, it's an experience I think I've been waiting for all my life, and—until now—I didn't really understand that it was possible."

Where the Sidewalk Ends

Given salamander-chasing's rigorous schedule, Mike and his wife, Lynn Siefferman, who is also a biology professor, trade off fieldwork nights. When she calls the next day to invite me to join a group of community members she's leading to a different pool, I can't imagine that she's going to direct me to a site more unexpected than the local gun range. But then she asks, "Do you know the place where the sidewalk ends?"

With no additional clues, I know exactly where we're headed. She isn't referencing Shel Silverstein's famous children's book; she's talking about a locale in a local park that has, for years, served as a turnaround point for walkers and bikers and

skateboarding kids. It is an abrupt ending that makes people pir-
ouette just before the sidewalk turns to grass.

I have always pondered why it ends there more than I've con-
templated what, exactly, lies beyond it. And what's beyond it,
Lynn tells me, is a particularly active ephemeral pool. She refers
to it as Barbwire Pond, due to the twisted metal pieces of agricultural fencing that divide the public park from private pasture.

The place where the sidewalk ends might hold an ephemeral
pool, but the place where it begins is flanked by soccer fields.
When I arrive, hours after sunset, it's ablaze with stadium lights,
though only a few people are kicking balls around. It looks like
the site of a spaceship landing.

Under the glare, I see Connie and Doug Hall, educators
locally known as the Frogologists. They're famous for lugging
amphibians around to schools so that students can commune
with them. "When I was into bird-watching, I used to always be
looking up," Connie says. "Now that I'm interested in salamanders, I'm always looking down."

She waves her hand across her face like a windshield wiper,
as if she might swipe the stadium light gone. "Out where we live,
there are nights when you can still put your hand in front of your
face and not see it. That's my preference. Sometimes, I'm so both-
ered by lights like these, I wish that I could just shoot them out,"
she says. Given my time at the gun club, this has real-world resonance. In contrast to that somewhat nail-biting setting, tonight,
I'm at what most people would consider a family-friendly place.
But the lights here, to some biologists, read like an acute threat.

We're soon joined by a group of diverse citizen scientists.
Together, we trace a sidewalk along the New River—and then
we step onto grass. Darkness thickens. We hear Barbwire Pond

before we see it, with a frog chorus so loud that, at close range, my ears start burning. Along the pond's bank, tiny spring peeper frogs shriek. And in back, wood frogs go *quack, ca-quack*.

With the help of headlamps, we can see across the pond to the base of a mountain. This land is protected by local government, which has installed hiking trails on it. "I don't think I've ever heard wood frogs before this week," I tell a retired biology professor, Wayne Van Devender, who's standing nearby. He's dubious. Maybe, he suggests, I just didn't know what I was hearing.

"It's like I used to tell my students: You really can't see anything until you know what you're looking at," he says. "Same thing for listening. The more you learn, the more you can see and hear. One thing's for sure—it's a good day to be a wood frog. Just look at all of them!"

But I can't see a single member of the quacking chorus. When I tell Wayne as much, he instructs me to put my diffuse flashlight beam—which has proven to not be very good at probing dark waters—directly next to my eyes. I stop shooting from the hip, as directed.

Wayne suggests using a fallen branch as a landmark. "Follow it along that bank to the edge of the water. You'll see wood frog eyes staring back at you." My flashlight catches on a string of silver beads, barely surfacing. Frog eyes, finally.

These creatures will only cluck for a few weeks, though peepers will sing longer. I thought I was unfamiliar with the song of wood frogs. But once I've met them, I start recognizing their voices everywhere. Wayne's right—I've just previously dismissed their distinctive voices as background noise.

I step over barbwire fencing to join a guy in his twenties who is crouching low to the ground, directly addressing an animal. I lean in to see that he's talking to a peeper frog, barely larger than a dime. "Could you be any cooler, bruh?" He turns his attention to me. "Just look at this little homie!"

On cue, the frog puffs to the size of a quarter. When he lets loose, the student puts his hands over his ears and shouts to be heard over the close-range belting: "It's louder than a Def Leppard concert out here, and I forgot my earplugs!"

"You know," Connie says, "in winter, wood frogs freeze solid."

"Like cryogenics?" I ask, channeling every sci-fi movie I've ever seen.

Connie nods. "They're fragile when they're like that," she says.

I'm reeling from the cryogenics revelation when one of the graduate students who's joined the group, Chloe Dorin, strolls over. Chatting with the Frogologists, she brings up the fact that spotted salamanders—animals who spend their entire adult lives in darkness—are the only known vertebrates on Earth who can photosynthesize, because they have algae living inside of their cells.

It sounds unbelievable. It sounds like science. Truth is stranger than fiction, almost always.

Naturalists discovered the relationship between spotted salamander eggs and algae in the 1900s. The globs of salamander eggs in this ephemeral pool will, in time, turn green. Because when a salamander embryo develops a nervous system, emerald-colored algae blooms inside of each egg, giving the salamanders a boost of oxygen, which the algae produce through photosynthesis.

It wasn't until the 2010s that researchers realized the algae seen inside of salamander eggs is also found in the cells of salamanders themselves, in an especially intimate interspecies alliance. It has since raised questions about similar relationships that remain undiscovered, given that this oddity was found in such a common salamander species.

Chloe says, "The more I learn about the salamanders' relationship to algae, the more questions I have. It's a one-of-a-kind relationship—a vertebrate animal that basically has plants living inside of it. That's some science fiction sort of stuff! Why does that occur? What are the ultimate effects? We've learned a lot, but there's a lot we don't know."

Salamanders' reduced capacity to self-identify bodily cells is thought to be related to their ability to grow new limbs and organs as adults. It's akin to the porousness that allows them to breathe through their skin, which makes them susceptible to taking in pollutants. Their physiological boundaries are unusually fluid, their sense of self larger than what we, as humans, might be able to comprehend. Salamanders, on a cellular level, seem to recognize themselves as part of a larger, living world—for better or worse, leading to sickness and to health—with vulnerabilities that often double as strengths.

It's not entirely understood how salamanders navigate to the same pool year after year as they do. It's thought that there's a strong magnetic-orientation element, some use of celestial navigation, and maybe even an olfactory aspect since salamanders can recognize the smell of home. Unfortunately, initial studies have shown that some of their navigational senses are disrupted by artificial light at night. Apparently, it's not only the light that confuses them; it's the resulting dark-light contrasts. Because artificial light leads to artificially dark shadows cast, and the

interplay creates a perceptual fencing on the ground that stands
to hold them back.

There has been a dearth of research on how artificial light
might hinder their regenerative abilities, in addition to their
navigational skills, but initial research out of Utica University
indicates that spotted salamanders' ability to regrow limbs might
be disrupted by it. Under artificial light, it is as if a salamander's
regenerative ability becomes self-consciousness—frozen, on a
cellular level.

~~~~~~

The rest of the group has left, but Chloe and I linger. I turn my
headlamp from white light to red. She makes her flashlight go
completely dark. Up ahead, the stadium lights have been turned
off, but security beams in empty parking lots are still forming pools of light on the ground below. Chloe tells me that the
migration might progress throughout the week as rain continues.

"For me, there's a deep feeling of awe in the migration," she
says. "It is so comforting, how elements in the natural world
confer to make this happen. The spotted salamanders we just
saw have been in those ponds before. They're coming home.
Some might have hatched there thirty years prior to come back
year after year. Their ancestors have been here for hundreds of
years, thousands of years, millions of years. I feel less alone seeing them. They make me feel like I don't have to fight all the
time. The battles aren't all lost. This is still happening."

There doesn't seem to be another human outdoors for as far
as we can see, which is, by the beams of security lights, nearly a
mile. After spending hours in the relaxing darkness of Barbwire
Pond, the lights strike me as particularly out of place.

In the 1990s, Russian scientists launched a fake moon into

space to increase human productivity in what was then known as the Soviet Union. The experiment planned to capture runaway sunlight to reflect it back onto Earth to disrupt the natural cycles of night. It worked. Briefly, the luminosity of roughly five moons illuminated landscapes in Europe before moving onto Russian soil. The experiment led to plans for whole fleets of reflectors that might allow the banishment of night at human will. Ultimately, the project was abandoned when some of the fake moons caught fire.

Though the idea sounds outlandish, one of the main arguments supporting fake-moon creation was cost savings. Proponents have suggested that, if sunlight is never allowed to fully take leave from Earth's surface, we will no longer have to pay for streetlights. Those fake moons were promoted not as a new world order but as a way of replacing the outdoor lamps we're already using. Futurist predictions of endless days are observable right now, in this parking lot, in my small Appalachian town, via a scene that's being replicated in parking lots across the country at this late hour.

A week ago, I would have driven by this artificially lit environ without giving it—or the ground under my car tires—a second thought. The parking lot is so bright that Chloe is forced to shield her eyes from the light that's pouring from above, though it's nearing midnight. "Why are these lights even here?" she says. "It's so weird, the way we live."

~~~~~~~~~~

Hoping to commune with fairy shrimp—tiny freshwater crustaceans—on my next ephemeral excursion, I reconnect with my old friend Wendy, trekking again from overlit parking lots to Barbwire Pond. Wendy, who lives a few miles away from

Barbwire Pond—in the center of downtown Boone—is one of the most enthusiastic and nature-knowledgeable people I know, and she always seems to be prepared. On the banks of Barbwire Pond, she pulls a recycled food container from her pocket and scoops water. Then she takes the concentrated light of a scuba diving flashlight and places it underneath so that the fairy shrimp she's caught are backlit.

I've loved fairy shrimp since the late 1980s—only back then, I didn't know it. They're related to the hybrid brine shrimp marketed as Sea-Monkeys in the back of comic books. Hungry for wonder, when I saw those advertisements as a child, I always wished I could buy some, though I never sent in a check or money order. Now I know that throughout the years I spent pining, I was likely surrounded by Sea-Monkey relatives without realizing it.

These fairy shrimp, barely larger than fireflies, are pale pink, with touches of citrus orange and purple trimming their translucent bodies. Threadlike appendages twinkle like fingers tickling a piano. "They look like little aliens," Wendy says.

She pours the fairy shrimp back into the pool as we're joined by her friend Leila and Leila's preteen daughter, Lou. "Have you heard of axolotls?" Lou asks, referring to the pink-skinned salamander species that's arguably the most famous in the world. I tell her that I have and that, just recently, I saw a kid wearing a T-shirt that said: *Axolotl Questions!* Lou giggles.

Axolotls sport feathery external gills that look like Las Vegas headdresses. Their likeness has been immortalized in video games and various toys. I have only just learned that juvenile spotted salamanders have gills very similar in shape to the ones that have helped make the axolotl famous. Yet for many kids at the elementary school located a mile from this place, their best hope of seeing spotted salamanders is via the Frogologists' mobile aquariums.

"Axolotls are huge!" Lou says. "They live in these lakes in Mexico." Their presence in popular culture does make them seem gigantic, but they're around the same size as spotted salamanders, roughly 10 inches long. And they do live in Mexico's lakes—but not many of them. Axolotls are thought to be the most populous salamander in the world. There might be as many as one million axolotls living in aquariums globally, but in Mexican lakes and canals, there are almost none left, due to environmental degradation.

Almost all pink axolotls in captivity are thought to be descended from a specific leucistic-mutated male that was shipped to Paris in 1863. Since that time, people the world over have bred axolotls for bubblegum-colored skin. But, in the wild, axolotls are almost always brown, animate extensions of their home soil. Pink axolotls are, by contrast, superstars of the human-built world.

In Mexico, the axolotl is famously associated with Xolotl, god of fire, who shape-shifted into a salamander to escape being sacrificed by humans. In regenerative medicine research, there's no species more famous. Researchers are looking to salamanders to solve all kinds of medical mysteries. The animals have already helped them understand spina bifida, and salamanders' resistance to cancer—which can be up to a thousand times more effective than that of mammals—makes researchers hopeful their bodily knowledge might someday transfer.

Leila leans into Barbwire Pond to take an underwater-camera photo of a spotted salamander, not-so-distant relative of the axolotl. "This almost feels like scuba diving. It's a whole different world in there!" she says. Only it isn't a different world—it's the one we've always belonged to, a living mystery.

Wendy points out a group of dragonfly larvae. Mature dragonflies might not look like dragons, but their larvae do, complete with armor. The silt-colored creatures move like hovercraft, propelling themselves through the pond by ejecting water from their rectal chambers with a pressure equivalent to that of a sports car. This was one of the first insects to evolve on Earth, 300 million years ago. And tonight, there's one swimming in Wendy's hand. "I call these butt breathers," she says as the larva shoots from her thumb to her pointer. At this, Lou snort-laughs.

All the times I've seen a dragonfly flit on blue-sky days, I've been witnessing ripples of dark places like this. I have, for all my life, been watching the marvels of darkness without recognizing night as the source of their beauty. I've been unaware of how a pilgrimage to a place like this might, after a hard, cold season, help soften my fragile spirit—though I now understand that it's exactly what I've been craving.

This ephemeral pool is like some sort of elixir that's potent enough to change my view of a place I thought I already knew. It is a concoction of hidden biodiversity that will ultimately populate the world around us—flashing through skies in the form of dragonflies, filling entire mountains with salamanders. Even when the water of this ephemeral pool evaporates, this spot will be covered in dehydrated fairy shrimp awaiting next spring's deluge so that they might reanimate in ephemeral pools the way those Sea-Monkeys were advertised as being revived by kitchen spigots. In a hushed voice, Leila says, "It feels like we're in on a secret, doesn't it?"

Watching a Salamander Dance

Before I started spending time with salamanders, I'd never considered that it might be easier to slide across rain-glazed ground. I hadn't thought about how moisture turns hard-edged leaves into tissue paper. I'd never looked up at bright clouds on a warm spring day and wished they would swell into dark forms. But that's what happens the following day. Every hour of sunshine makes me think to myself: *I wish it would just rain already!*

When I tell Wendy that I'm headed out again that night even though clouds have failed to gather and temperatures seem below ideal, she offers to accompany me. Neither of us consider ourselves big news-watchers, but recent events have been acutely concerning. The specifics of doom are ever-changing. What remains the same is how scrolling newsfeeds makes me feel like my soul is shriveling.

My headlamp has, night after night, proven too diffuse for probing water. It is more of a bludgeoning device than a scalpel, so Wendy has brought her husband's scuba light for me to borrow. I'm happy to have a better tool, but I'm having trouble figuring it out. The flashlight doesn't have a switch; instead, its casing requires a twist. And I've twisted it too far, all the way to a distress setting that looks like the reflections of a disco ball. As the light *thump-thumps* against my calf, Wendy laughs. "How are you even doing that?" she asks.

The nightclub-style strobes inspire me to tell her that one of the reasons I've been excited to witness the migration is because I've heard that spotted salamanders perform artful courtship rituals. They nudge each other and move in circles. The activity is so rare that some field biologists who've gone out for years have missed witnessing it. Still, I'm here because, against all odds, I want to watch a salamander dance.

Wendy gets it. After a hard winter, she has specific natural-world yearnings of her own. "This time of year, I just want to get my hands dirty," she says. "It's the best way I know to combat the winter blues."

It's well documented that bacterium in soil can boost moods, maybe as effectively as antidepressant drugs. While light therapy plays a role in increasing serotonin and has long been used to combat seasonal affective disorder with morning exposure, researchers have started sounding warnings that being exposed to artificial light at night warrants more attention for its contribution to rising rates of mood disorders.

It's been linked to pro-depressive behaviors, and it activates the part of the brain associated with disappointment and dissatisfaction. Brains that process artificial light at night are known to have lower dopamine production the following morning. But, given our cultural associations, it's difficult to align with the concept of light as a force with the power to make us sad. We almost always use darkness to symbolize depression and light to indicate happiness.

Parents—including me, mother of a twelve-year-old—are increasingly concerned about how often their children are exposed to screens, but a large-scale study put out by the National Institutes of Health in 2020 shows that other forms of artificial light stand to cause harm. Adolescents who live in neighborhoods with high levels of outdoor artificial light at night are more likely to suffer from mood and anxiety disorders than those who live in areas that still have access to natural night. Rates of bipolar disorders and phobias in particular have been found to rise with lighting levels.

The human relationship to artificial light is relatively new, but our relationship to natural darkness is ancient. This seems obvious, but it's hard to absorb that, unlike society's most prevalent light-dark metaphors, light is not always a positive force and darkness is not always a negative one.

It's Friday, but there are no stadium lights tonight. The soccer fields are quiet. It hasn't rained, and there's nary a worm on the sidewalk. "I don't think we're going to see anything," Wendy says, directing our attention across the New River, where security lights reveal the contour of a distant riverbank.

Instinctively, we turn to identifying the lights as if they are stars forming constellations. There's a car dealership. A produce-distribution center. A new indoor gym that Wendy sighs at on sight. "That's right on the river, one of the prettiest places in the world, and they didn't put a single window on the backside of that building," she says. Even in full daylight, people on treadmills cannot see the river running alongside them.

We recognize another cluster of lights as a gas station. We're temporarily stumped by a parking lot that appears to be the brightest spot in the lineup. It's directly across from the ephemeral pool. Ultimately, Wendy identifies the lot as being attached to the administrative offices of a local electric company. She shrugs.

"I guess they were like, 'Well, we've got the energy, might as well use it!'"

Outdoor lighting at night often gives people a sense of security, but there is not clear scientific evidence that it increases safety. In fact, some studies have found that streetlights do not lessen accidents or crime. Certain forms of security lighting have even been found to decrease safety since they make potential victims and property that might be stolen or vandalized easier for perpetrators to visually target. It's a fraught topic with no easy answers, but the fact remains: It's not uncommon for outdoor lights to be installed haphazardly, favoring as much illumination as possible with little thought as to how darkness might situationally be of aid.

Temperatures keep dropping. Our breath appears as chalky puffs against a blackboard. We walk back toward the ephemeral pool, drawn by the promise of life. At the place where the sidewalk ends, we crouch in foliage that shields us from the lights across the river. In my scuba beam, I can see fallen twigs with salamander egg clutches attached to them. And there, in the middle of the pond, embracing one of those clutches, is a female spotted salamander. The egg mass she's clinging to resembles NASA images of a star expanding. And right now, she's adding to it.

"She's laying eggs! Right now!" Wendy says. She grabs my arm in an I-can't-believe-this move, and we jostle up and down like we've just won the lottery.

Next to the egg-laying salamander, the pale flash of a second salamander's belly appears. Then it disappears. This isn't the courtship ritual I've read about, in which males attempt to woo females by nuzzling their heads, but it's clearly some type of dance. "Did you see that?" I ask Wendy. But she is still hyper-focused on the egg laying.

"I can't believe this is happening! We need to be careful that we don't disturb her," Wendy says, turning her scuba light from its high to low setting.

Meanwhile, two frogs locked in an embrace pop to the surface of Barbwire Pond. They're surrounded by a halo of fairy shrimp who are fanning their pink-feather legs, flashing tropical colors in dark waters. Through the cloud of translucent fairies, the frogs swerve into a gelatinous mass of already-laid frog eggs, shining purple. Their movement is making everything quiver.

That's when I see the dancing salamander again. From the darkest part of this pool, in the darkest part of this beloved park, the salamander's star-dotted body shoots again to the surface of water. This time, Wendy sees him, too. "Oh, my!" she exclaims.

The animal is flipping. He's somersaulting. His feet-hands are fervently waving back and forth, churning with such joy that we start laughing along with the wood frogs as, all around us, the peeper frogs howl. Every species in this ephemeral pool seems to have come alive at the same time. The water is writhing with life. It twists and turns, colors of a kaleidoscope, until I've lost my bearings.

When the frogs kick off into deeper water, it feels as though they've conjured a celestial wind, knocking us out of the hypnotic state we've been caught in. Wendy and I both take a step back at the same time and gasp in unison. It's that tangible, the feeling that we've just surfaced from swimming in a pool of life everlasting.

What we're feeling is awe, which the *Journal of Positive Psychology* defines as "a self-expansive emotion, where the boundaries of a separate self are transcended to process a larger, complex reality." Neuroscientists have found that, when humans experience awe, it potentially reduces the brain's capacity to read bodily boundaries.

In awe, the egocentric sections of the human brain appear, on scans, to go dark as the organ moves to process "you and me" into "we." In this way, awe isn't just an emotion; it's a physiological experience that temporarily lessens our neurological firing. Awe is, in this sense, a resting state.

"Wow," Wendy says. "I feel so full."

I, too, feel satiated. Like my soul has been rehydrated.

We came to Barbwire Pond reeling. We are still wrestling with the reality that justice doesn't always prevail. We are reckoning with the fact that, even in the modern world, safety is often an illusion. But life on Earth is tenacious. And we are part of it, indivisibly.

The term "mucked up" is used to indicate that something is wrong. But muck is life-giving and life-sustaining. I have the urge to tell everyone I know about this place, like an ephemeral pool evangelist. But trampling human feet would imperil the very ecology that deserves attention. Still, traffic to this singular pool of wonder need not be dangerously concentrated.

There are, around the world, tens of thousands of ephemeral pools. Maybe millions. The reality is: Nobody knows. Ephemeral pools have been found all over this planet. They form a string of earthen bowls from Portland, Oregon, to Portland, Maine. There are some citizens taking inventory of pools in states like Vermont. But they mostly exist without humans noticing.

They're notoriously hard to find because, much of the year, they aren't there. Yet creatures like fairy shrimp remain in their absence, waiting to be reborn with rain. A single ephemeral pool, like this one, can hold more biodiversity than its fifty surrounding acres. On this imperiled planet, magic still exists, often in the places we least expect it.

This ephemeral pool is, like almost every ephemeral pool, threatened. A municipal complex, with another parking lot full of artificial lights, might soon be built across the river. This sidewalk ends abruptly because someone has plans to extend it. But Barbwire Pond is right here, right now. There are thousands of people who walk this path in daylight, turning around where the sidewalk ends. They don't know this exists. And it's hard to protect that which you haven't really seen or heard or been given the opportunity to consider as important.

I've been examining the surface of Earth at close range. And all the while, I've been balancing on the balls of my feet, straining to hold my body aloft, somehow separate. I've been holding myself at bay while searching for animals on the ground, because—as ridiculous as it sounds to admit now—I didn't want to get my pants dirty. But the exhaustion of late nights has accumulated. The position I've been holding is not one I can sustain for much longer.

When Wendy walks off, I lower myself to the ground, rolling from my feet to kneel on soil that's still saturated from earlier storms. The ground is soft. It is forgiving. The relief of resting against it is somehow surprising. This is the respite I have been denying myself by not getting fully down and dirty, night after night. Because, despite how intently I've been focusing on it, the ground has felt somehow beneath me.

Until now, I have not really been aware of this underlying perception. I thought the pajama-wearing student's reaction to a salamander stooping so low as to walk around on the *ground* was sort of funny. But it's a deep-seated notion that I myself have been carrying subconsciously.

Before I started paying attention to salamanders, I never realized that there's peace to be made with darkness below as well as above. But now here I am, on my knees, ready to accept encompassing darkness—here, on the surface of Earth—as its own sort of savior. This might, to some, sound like blasphemy. To me, it feels not unlike an old-time Appalachian baptismal ceremony.

Even though it's now common for baptisms to take place in fiberglass pools and galvanized cattle troughs full of chlorinated water, it isn't unheard-of to come across more traditional ceremonies at municipal parks like this one in summer. I have, with my own eyes, seen people go down to the river to pray in that old way. I have watched entire Baptist congregations gather along riverbanks in white clothes to be dunked in the flow of water that has not been sanitized—that is to say, water that's still full of life-forms, including some of the largest salamanders on this planet: hellbenders.

Those giant Appalachian salamanders, which live in the river just beyond Barbwire Pond, can grow to be nearly three feet long and were named such because the European settlers who saw them thought that they looked like they came from hell and that when they twisted themselves under river stone, back to hell they went. But in that river, where parishioners are dunked to symbolically shed sin in baptism, their white clothing sometimes gets soiled, taking on the color of silt, the color of salamander skin, the color of local riverbeds.

Here, at the place where the sidewalk ends, I have found a shared sanctuary. And in it, I have gotten thoroughly cold and wet and dirty. Clearly, this wouldn't be everyone's idea of fun. I'm not sure it's even mine. But in the muck—with salamanders, fairies, and dragons—I am coming to understand, in a visceral way, that being grounded is not a punishment; it is a privilege. The dark recesses of this planet are not hell; they are home.

Tonight, the houses on the ridgeline above the park look like aquariums that have been filled with light. But down here in the valley, I feel as though I've been poured from my own container back into the whole. Down here, fairies are tickling my fingers with their dancing feet, reminding me to pay attention.

Darkness, like a mountain-surfing salamander, migrates toward us every day, sliding across the burning face of this world. But welcoming it would be countercultural. Artificial light is a consumer good, bought and sold, while natural darkness offers itself freely. It is a renewable, life-giving-and-sustaining respite that we, as humans, are increasingly depriving ourselves of because we have been told, so many times, in so many ways, to fight the dying of light. But at Barbwire Pond, it's clear that surrendering to natural night would be a communal victory.

The same darkness that presses against celestial bodies will brush against our faces, cool as a mother's hand against a fevered cheek, as soon as we let it. As soon as we, like brigades of crossing guards, hold space for it. I'm only visiting this spot temporarily, like most of the migratory creatures in this bubbling pool. I'm surrounded by lives beginning and continuing, cycles new and ancient. This is but a way station in a never-ending migration. But right now, there's a salamander dancing in the darkness that remains. And to bear witness, I only need to be still. I only need to be.

I give the end of my scuba light a gentle twist, skipping the disco setting for a slow fade. As light slips away, my contracted pupils begin to widen to absorb as much of this place as I can. Leaves, branches, whole trees are swept up in the incoming tide of darkness. Gone are the threads of silver barbwire, the delineation between the place where the sidewalk ends and the one where life begins. It is difficult to even see the outline of my own hands. My arms, my legs, all my body parts dissolve as the inky borders of the ephemeral pool expand to include me.

Owls Nesting

~~~~~~

The Astronomer and the Owl

On one of the very nights that Wendy and I were probing the dark in search of spotted salamanders, a streetlight went on in her downtown Boone neighborhood. Its bulb had been broken for years, creating a snaggletooth gap in the bright smile of lights that curve around her block. In a lot of places, a dormant streetlight's activation would be seen as cause for celebration. But in her neighborhood it was viewed as a life-or-death concern demanding immediate attention.

Her neighbor a few doors down, Richard Gray, professional astronomer, was the first to notice the change. This seems fitting, since an astronomer's job is, at root, to study light. Astronomers take images of nebulae, swirls of dark and bright, and interpret what's happening in the interplay. As a researcher, Richard uses spectroscopes to break stellar light into component colors the way prisms reveal rainbows, with hues suggesting a star's age and velocity.

Years ago, he had to travel winding mountain roads to access the telescopes at Appalachian State University's Dark Sky Observatory, located several miles past my own neighborhood, which is on the outskirts of Boone. Now he just logs on to apps from his computer, which gives him the power to remotely access telescopes from his house. The whole endeavor could be handled at a distance via screens and buttons, but he still often steps into his yard beforehand to gauge local weather.

It was during one of these regular surveys that he noticed a streetlight-maintenance truck. In recent years, the Town of Boone has phased out traditional lights for more energy-efficient bulbs. County workers weren't aware that this streetlight had been nonfunctional for years; they were just going about their assignment.

The additional artificial light would have, under any circumstances, been worrisome for a man who spends his life attempting to decipher starlight. But Richard was doubly worried, because the light emitted wasn't only interrupting his stargazing—it was reaching into a wooded lot across the street, all the way to his newly installed owl-nesting box.

For the last few years, Richard and his wife had enjoyed a screech owl singing them to sleep, and he had just built a nesting box in appreciation. As a lifelong birder, he knew that screech owls don't craft nests of twigs and twine. They're cavity nesters who tend to find already established holes in trees. So in most places where mature trees have fallen to development, owl boxes can be helpful.

After Richard put the box up, he'd waited for owls to take him up on the offer of shelter. And he'd waited—and waited. Then, just before that streetlight went back on, a screech owl pair had shown up. It's common practice for owls to house hunt before settling down, so the light had particularly bad timing. The relationship he'd worked to establish was acutely threatened.

The new bulb was, like almost all newly installed bulbs, a white-light-emitting diode (LED). And, like many LED lights, it produced blue-rich white light. This type of light isn't just the most damaging for astronomers' sight; it's one of the most biologically confusing for wildlife, since animals read blue light as a signal to enact daytime behaviors.

Astronomers tend to think of light not in generalities, but by wavelength: Red waves are long and easygoing, blue are short and energetic. We might dismiss sunlight as a generic white, but it's violet and blue and green and yellow and orange and red. We're surrounded by rainbows every day, we just can't always make out light's diversity—mainly because blue wavelengths are particularly pushy. Given their density and size, they tend to scatter more easily than others, which is why we see sunny-day skies as blue.

When animals are blasted with blue light at night, their bodies naturally read it as daylight. This is why, with an increase in light pollution, biological systems three billion years in the making are starting to misalign. Animals get amped for their respective activities out of sync. Artificial light literally puts us on different wavelengths.

In contrast, warm-tone spectra tend to be read by animal bodies for what they are: more easygoing energy. Campfires and old-fashioned incandescent bulbs tend to pool light where they're located. But blue wavelengths behave like beads hitting hardwood floors. Even in situations where streetlights point downward, as responsible guides suggest, blue light tends to bounce skyward.

Advanced LED bulbs provide more light with less energy. This has led to near-universal adoption of blue-rich LEDs, though the technology is capable of other colors. Life scientists generally agree that LEDs with high blue content have larger biological impact than the bulbs that came before them. And the potential energy conservation of using this technology has largely been negated by human behavior. Instead of using less energy after switching to LED lights, people have, the world over, started using even more energy for ever-brighter illumination, as if we've collectively decided that, since the light is energy-efficient, we might as well produce more of it.

Milan, Italy, was one of the first European cities to convert to white LEDs. By 2019, more than half of the city's streetlights had been switched from warm-tone bulbs. The trend is observable around much of Europe, and North America's light pollution is increasing at an even faster rate. In 2023, the U.S. Department of Energy instituted regulations to halt production of the 30 percent of incandescent and halogen bulbs still being offered, limiting warm-color options. They reported that the switch would reduce carbon emissions by 222 million metric tons over 30 years, the equivalent of the emissions generated by 28 million homes annually.

They stated that this could save the average American family $100 a year, but there are no clear estimates on the biological costs of increased blue light—not to humans, not to owls, not to any living being on this photosensitive planet. In fact, there is a dearth of data on the toll it has already taken or even how much artificial light we're engulfed in after sunset.

Many news stories about light pollution present it as having increased roughly 50 percent over the past 25 years. But people familiar with satellite technology have brought to attention the fact that some of the equipment that's been used to gather data is not sensitive to blue wavelengths, which are on their way to global domination. With blue-rich LEDs taken into consideration, some scientists suggest that, in certain places, artificial light pollution has more realistically increased 270 to 400 percent during the past 25 years.

Interestingly, as outdoor blue light has proliferated, literally under radar, our awareness of it indoors has increased. During the pandemic, when nearly every interaction with people outside of a given household—be it personal, professional, or telehealth—was conducted via screens, the popularity of glasses with blue-blocking lenses grew. We know blue light is harmful, but it seems impossible to get away from it. After all, when Richard stepped away from his computer on that first LED-lit night, he found that the same blue spectra that he'd been trying to take a break from had swallowed his entire house from the outside.

He'd planned to watch the miracle of owls' seldom-observed life cycles unfold with the same close attention that he gives young stars through telescopes at the observatory. Now, every tender act he'd hoped to witness was under siege—along with the well-being of every living creature on his street.

Frantic to find an off switch, he started making calls. To public officials. To neighbors who might help him create an uproar. But in the end, no uproar was needed. The town hadn't known the owls were there. Once administrators learned that a brooding pair might be disturbed by the streetlight—which they'd thought, if anything, human residents would consider an upgrade—workers arrived within seventy-four hours to remove the bulb, returning the neighborhood to the level of darkness it had previously enjoyed.

Municipal administrators had been trying to do the right thing with those LEDs. They had crunched facts and figures, but they had not absorbed the true cost of pelting ice-hard energy against every spring-soft creature in its radius. Unfortunately, the town's quick response was still too late. By the time the streetlight went dark, those house-hunting owls had already fled.

Blue Light Refugees

No one can say for sure that artificial light was the only reason the owls left. But what the Cooper family can confirm is that—a few doors down from Richard's house—the same LED light that hit his owl box flooded their living room. It pushed through curtain cracks to cover their children's beds. Immediately, they decided to invest in blackout blinds to guard against light. Otherwise, the carefully scheduled equilibrium of their household—with two energetic kids and a rambunctious dog—was going to be thrown off.

Outdoors, the wall of their wood-sided house, maybe 200 feet from Richard's owl-nesting box, acted as a dam that stopped the highest tides of LED-light from reaching the forested lot on the other side of it, as well as a small space between their house and the one behind it. It was in that surviving woodland, roughly three trees wide, that they found a blue-light refugee peeking out of a hole in their chestnut oak tree.

The hole was one they kept an eye on during nesting season

because, for years, the oak had been home to a family of nut-hatches. But hosting screech owls was a special privilege. Soon, neighbors were stopping by to get a peek, whispering about their observations at low volume, since owls are particularly sensitive to noise, as well as artificial lighting.

Sarah Cooper, an old childhood friend who'd heard about my nocturnal-creature communing, started texting me photos of the female bird, framed by the oak hole at sunset. Around the same time, another friend, who lives in the house behind Sarah's, began sending images of the male, who'd started napping in a moss-coated tree beside her second-story balcony. Before long, my text message threads made it seem like I was part of a vast network of owl-tabloid reporters.

Ultimately, Sarah invited me over to see the owls in person. She explained that, during nesting season, a female screech owl waits in the tree all day. But when she senses that darkness is deep enough to ease her mobility, that mother bird launches into the wider world to take a break from acting as sentry, giving the owlets' father a chance to tend to them. Sarah suggests that, though she has not seen any owlets, she's sure they've already been born. Because, when the wind is just right, it sounds like the tree is singing.

~~~~~~~~~~

It's near dusk when I meet the screech owl. I am alone since the Coopers had to go out. But, as promised, they've left me a monocle scope. I find it sitting on their deck, already trained on the oak, its trunk barely a hundred feet from their back door. Without the already focused device, I doubt I would have been able to find the nest hole—or the small, feathered face filling it.

The owl's feathers are indistinguishable from the bark of the tree she inhabits, a mottling of browns and whites with hints of copper. Her wide, mossy eyes match the green of early spring leaves. In the scope's frame, the owl blinks. I blink. I am watching her. She is watching me.

In many cultures, looking into owl eyes is ill-advised. Owls are, around the world, sometimes seen as symbols of luck, but they're just as often representative of witchery and death. Regardless of cultural connotations, a recurring storyline throughout human history has been that these birds are messengers with the ability to travel between earthly and otherworldly realms.

Even among seasoned bird-watchers, finding an owl nest is considered a holy grail. Owls, like many nocturnal animals, are under-studied, not in small part due to how hard it is for researchers to scour the wild after dark. Raptor centers repeatedly suggest that no one should reveal the location of an owl nest, lest the place be overrun by curious onlookers. I'm honored that this secret spot has been shared.

I can hear bits of conversation coming from a house through trees. Someone on a porch another road over sneezes. Cars are barreling by. On a mountain-cut roadside above me, a delivery truck hits a bump, metal doors rattling. Finally, things settle.

Coo-ah, a bird calls from somewhere nearby. *Coo-coo-coo*. Or is that *who-who-who*?

I straighten my spine. Close my eyes. Focus.

At first, I think this is the mate of the owl in the nest, the father who is somewhere close by, waiting for dusk. But I soon realize that I'm listening to a mourning dove. Its cooing sounds like the owls of cartoons, the *who, who* of children's books, not the lilting call of the screech species. Frankly, the dove I'm hearing sounds more like a stereotypical owl than the actual owl recordings I've listened to in preparation.

When a motorcycle rumbles by, I see the noise register in the mother owl's body as a slight neck twist. The human driver is wholly unaware that his machine is disturbing an unseen nest. Above, simultaneously, there is the rumble of a small aircraft, the sort that someone might hire for a mountain-view tour. But we are shrouded under shade trees.

The Coopers' house is already in shadows due to the surrounding, layered mountains. The gradual rise of darkness carries with it a roiling owl song. It's not screeching in the least. It's more of a trembling pony whinny: *Weheeheehee, weee.*

The owl is letting his mate know that he's nearby. It is a call that screech owls make to keep in touch with their families when they are apart: mother in nest, father somewhere around, tucked into the nook of a branch. Mated screech owls have been known to sing to each other day and night.

These owls' musical thread, connecting them despite their distance, evokes time spent with my own elderly family members through glass during pandemic years. It makes me think of the doors set between my loved ones during quarantine. Porch visits with friends, metal rails that held people out and in. The close-yet-so-far trilling makes me think of how people banged pots and pans from high-rise windows in cities. How, while all those things were happening, people reported a heightened awareness of birds, noticed other species calling out their own songs of love and mating and warning. We were together. We were alone. We are still surrounded by nonhuman neighbors who chat and enact dramas of their own.

Gnats swarm the owl's eyes. She blinks and blinks before

retreating into the tree, leaving me to stare at a black hole. My eyes turn toward the ground, which undoubtedly houses salamanders. I have played with toddlers in this woodland myself. Those days—spent with my son and his friends, including Sarah's kids—reminded me of how a wooded lot, or even a few trees, can seem like a vast wilderness to children.

When the owl returns, she seems to be taking a power nap before dad begins his shift. Light drains as she rests. *I am here*, her mate calls in quivers of assurance: *I am here, I am here.* He whinnies often so that she cannot forget.

I sit, without distraction, until the golden-sunset hour fades into gray. Until the entire scene has transformed to shades of tree bark. *Coo*, calls the mourning dove, reminding me that, in a very short span of time, I've begun to decipher languages that are not my own.

There is anticipation in the owl's stance, a nearly imperceptible inching. I do not move my eye from the scope. I want to catch the decisive moment when darkness hits just the right tone, a signal that only the owl can interpret. Finally, the oak tree's feather-camouflaged midsection swells like a tree burl, until, in a single swift motion, the owl flings herself into the sky, released by the blessed arrival of night.

~~~~~~~~

When I return to the Coopers' house a few evenings later, I find Pace Cooper, elementary science teacher and dad extraordinaire, having his short brown hair brushed by his youngest daughter. "I'm being preened," he says, laughing before he pauses to explain to his daughter that preening is how birds groom, pulling dirt off feathers with their tiny beaks.

Sarah—a veterinarian by trade—is standing at the kitchen counter, washing dishes. I don't see her often, but I've known her almost all my life. We went to summer camp together at a now-shuttered local establishment. We're both in Boone because, independently, we decided that we wanted to live at camp forever. Once, we were kids bunking in cabins named after John Steinbeck novels. Now, we're adults whose only evening plan is to observe owls.

When Sarah scoops up her daughter for nighttime routines, Pace and I step into the night. It's already too dark to use the scope. All we have are naked eyes and open ears. And we feel like that's okay. Owls aren't meant to be seen by humans so much as heard, anyway.

On our way out the door, Pace hands me a blanket from a pile reserved for campfires. In these mountains, nesting season can be chilly. We settle into patio chairs, and he tells me about how he used to work at a local wildlife rehabilitation center that often took in injured screech owls, sometimes even saw-whet owls.

Saw-whets, known as one of the most charismatic owl species, first came to my attention—and that of millions of other people—in the depths of the pandemic, when a worker decorating the 2020 Rockefeller Center Christmas tree noticed golden eyes peering at him from the interior of an 11-ton Norway spruce that had been trucked in from upstate. The worker thought he'd found a baby, given that the adult bird was barely seven inches in length.

The owl, named Rocky, became an internet sensation. Picture books were written. Songs were sung. Headlines appeared in the *New York Times* and the *Washington Post*. NBC declared Rocky to be a "bright spot" in a terribly dark time. But it was from darkness he came, and to darkness he would go again—and that wasn't a bad thing. Still, the media generally overlooked the fact that darkness is required for any owl to have a happy ending.

When Rocky was released into the wild, an Audubon Society owl expert suggested that, since the limbs of his former home had been covered in 700,000 LED lights, he would likely set out for Appalachia's sky islands—high elevation, cold-weather habitats of spruce and fir that are remnants from the Pleistocene epoch. These high-peak habitats—left behind when most boreal forests shifted during historic climate change—are isolated from similar ecosystems by hundreds of lower-elevation miles.

Sky islands are naturally darker than the land around them, identifiable as ink blots on light pollution maps, given that their peaks are too craggy for easy development—though they still need protection. Climate change from human activity—which is occurring ten times faster than the average rate of warming that formed these remnant habitats after an ice age—is rapidly encroaching, with warm air currents degrading sky-island habitat the way rising oceans eat away coastlines. But for now, the cloud-towering sky islands remain somewhat intact, surviving against all odds, like Rocky.

When I tell Pace about how Rocky captured my imagination, he nods, as if becoming obsessed with tiny owls on remote sky islands is a normal preoccupation. Before he dedicated himself to teaching, he was a fly-fishing guide. But before that, he was a biology student studying owls. Saw-whets, specifically.

I've known Pace for years, but it is only now, within listening distance of an owl nest, that I learn his graduate-school work included fitting saw-whet owls with tracking devices in the 1990s. He tells me that one night, an owl he was tracking went dormant for so long on Grandfather Mountain that Pace became nervous. Ultimately, he found the animal shrouded inside a tree. Determined, Pace went to a nearby hardware store to buy a ladder, and, using an angled dental mirror, discovered that the bird wasn't dead; she was nesting.

His discovery played a role in confirming that saw-whets are permanent Southern Appalachian residents, with some nesting-as-opposed-to-migratory populations surviving here, in remnant northern-species forests that remained when lowland evergreens were lost eons ago. Now, here we are, sitting on his patio, watching a screech owl with a saw-whet resemblance nesting with Pace's own family, each species keeping their distance.

This owl, likely artificial-light refugee, went from an astronomer's owl-box straight to the shadow of a house inhabited by two owl-loving life scientists. It is, like the story of Rocky, like something out of an avian fairy tale. "You know, I used to find saw-whets in Christmas tree farms all the time," Pace says of his tracking days. The tiny owls often use the farms as way stations as they move from sky island to sky island, across Appalachia, the way monarch butterflies use milkweed plants to support their fantastic migrations, though pesticide use is a concern for all involved.

Our county is full of tree farms. I live, as the crow flies, less than a mile from one. It's novel to think of them as way stations by which a saw-whet owl with evergreen needs might traverse the East Coast. But even during daylight, owls need darkness in the form of shade, just as trout depend on overhanging trees to cool mountain streams.

There very well might have been saw-whets visiting my own neighborhood while I was scrolling social media posts about Rocky. And there are surely screech owls nesting in woodlands near my house. I'd been thinking of sky islands as enchanted places at high elevation. But thanks to this tree-farm informa-tion, I'm now imagining the land between the Cooper house and mine as owl territory.

Above our heads, the mother screech owl is keeping watch, waiting for darkness to cue the father's temporal caretaking duties while she perches nearby so that he can easily deliver food to her as well as the owlets. Above us, the male owl whinnies. "Haunting, right?" Pace says. "That's why people use screech owl sounds in horror movies, to set up for dark situations. Their call is kind of terrifying."

When light fades to a pale gloom that's visible through the overstory, the mother bird leaves her post with the same trium-phant leap as before. Around the same time, Sarah steps out to join us, her own children tucked into bed. She clicks a propane fireplace on to supplement the warmth of blankets. Then, almost immediately, she turns it off. It isn't so much that she thinks firelight would bother the owls; it's because the stove makes a hissing sound. "This is way too loud," she says.

In the quiet that follows, she asks, "Hear that?" I close my eyes and hear the owlets chirp-chirping. When I open my eyes again, a songbird-size shadow smears the twilight above me.

"That's the dad," Pace says. "He just dropped off a snack."

Again and again, the male makes his way over and under tree limbs. Weaving and bobbing. Feeding and hunting. In time, the flying gets wild, with the owl swooping lower. "One just flew three feet over your head," Sarah says as we enjoy the aerial show. Owl parents move in and away, in and away. And, all the while, through the oak's open-mouth hole, tiny voices keep tum-bling out as baby birds practice their singing. "This is so cool," Sarah says. "The babies are louder this week."

"You know, our eyes would be the size of softballs if they

were proportionally the same size as owls' eyes," Pace says. Sarah laughs. She has the most striking eyes of anyone I know, crystalline blue framed by black-gone-gray hair, though right now, it's too dark to see any of this.

Owls' famed night vision employs the same rod and cone systems as human eyes, though they have a greater ratio of rods to cones. Their eyes also dilate to extreme extents to let in available light. They're more open to artificial light, so it stands to acutely harm them. At extreme levels, it can damage their retinas, going beyond hampering sight to permanently limiting vision.

The neighborhood is so quiet that I can hear the clicking of talons against the tree-hole rim when owls land. "It's usually the dad that takes care of feeding, but both parents seem to be at it tonight. A hundred and fifty grams, that's what a screech owl weighs," Pace says, in teacher mode. "It's like the equivalent of thirty nickels in your hand."

Birders commonly compare animal weight to nickels, but this inspires Sarah to turn on a flashlight so that she might show me an actual coin, gold that's been pressed to fit the curve of her finger. The ring, bearing the outline of an owl, had been an unexpected gift from her aunt long before these owls moved in. She'd worn it for months before she encountered a stranger—on an airplane, in flight, of all places—who had the same owl image tattooed on her arm.

"She told me that the owl is modeled after a famous ancient Greek coin," Sarah recalls. And, with that information, a portal to understanding owls' place in Western culture opened for her. The owl on the fifth-century coin Sarah wears is known as the Athenian owl. Sarah says, "Reproductions of the coins are more common than you might think." There seems to be market demand for them, owl lovers aplenty.

Ever since I started thinking about saw-whets living in Appalachia's sky islands, ever since I first locked eyes with this resident owl, I've been wondering: Why is it that we think of owls as smarter than the average bird? How did they come to be considered wise? In western culture, this traces to Athena, ancient goddess of wisdom and courage.

In Greek mythology, the owl rides through life on Athena's shoulder. The bird, as a nocturnal creature with supreme night vision, allows Athena to view truths that she could not see on her own. In this sense, the Athenian owl is a counterpart to human knowledge, embodiment of what we struggle to access. It is, according to Greek mythology, through tiny owls that darkness communicates its teachings.

The city of Athens, Greece, is thought to have been named for Athena not necessarily because of the goddess's supernatural prowess but because, in ancient times, tiny birds—a small species known as *Athene noctua*, resembling both saw-whets and screech owls—were present in overwhelming numbers. There is a still-used adage in Greece that speaks to this history. To needlessly give someone something that they already possess in spades is said to be like giving owls to Athenians. Even today—in a city once lit by torches that can now be seen as a blob of LED light from space—owls of *Athene noctua* lineage nest in the ruins of Athens's ancient darkness.

On a street below the Cooper house, someone walks by whistling a tune. Then the neighborhood goes quiet. "Hear them getting stronger now?" Sarah says of the unseen owlets. "They're up there just chit-chittering." It is a high-pitched, melodic sound.

Their mother is still away. Their father is off in darkness unknown. We imagine that they are crying, calling out for food, wanting to absorb the heat of their mother's body.

"Oh, Leigh Ann," Sarah says in a soothing voice. "Stay calm. Don't react." Her eyes are trained on something just over my shoulder. "Turn around really, really slowly. There's an owl right behind you. I mean, it's really, really close."

Anyone who has spent time around me knows that staying calm is not my strong suit. I am highly excitable, for better and worse. But it feels like the serene overtones of this natural night that's absorbed us might help my body match the tranquility of the moment.

Slowly, slowly, I turn. And on a rhododendron branch, barely an arm's length from my body, I see the dark outline of an owl. She is staring at me. I am staring at her.

I was at an advantage during our last meeting, in lingering daylight, but she's at an advantage here, in the deeper end of evening. Her giant eyes are taking in more moonlight than mine. Despite the bird of prey's proximity and razor talons, the encounter is not intimidating. It feels like a wild-world blessing. We have chosen darkness, and this owl has, in turn, chosen us. Across the dormant fireplace, Sarah swears that, from where she's sitting, it looks like the tiny owl is perched directly on my shoulder.

What Is Darkness, Really?

Richard, the owl-loving astronomer, has a gray wizard's beard. He greets me at his front door, where I notice that he has propped open the metal lid of his letter box with a clothespin. In its dark innards, a songbird has already woven a nest. Richard's mail carrier has been directed to leave any human missives on the stoop to avoid disturbing it.

I've dropped by because I'm curious about how an astronomer, professional interpreter of light, thinks about darkness. Graciously entertaining the notion, Richard leads me to his backyard garden, where we sit in the shade of a ginkgo tree. There I ask, point-blank: "What is light, really?"

"Light is energy traveling at us," he says. "That's pretty easy to explain: particles of energy, photons, propelled through space." Darkness, however, is a little more complicated. "Darkness is what you get in the absence of visible light," he says.

"But wouldn't that mean light is what you get in the absence of visible darkness?" I say, standing up for darkness, underdog of the ages.

Richard shakes his head. "Not exactly," he says.

I tell him that at the camp Sarah and I attended as children, we were taken into caves with carbide lanterns. In the subterranean depths of these hills, counselors would, at some point in the journey, always instruct us to turn off our lanterns to be introduced to what they called "true darkness." But, though it had appeared to be true darkness to me, I'd still been surrounded by light waves.

Richard explains that everything I've ever encountered has surrounded me with light, visible or at wavelengths beyond my capacity to see. Natural light is never absent, only present in states heightened or lessened. "Without temperatures of absolute zero, there can be no absolute darkness," he tells me. Absolute zero—the calculation of lowest possible energy or thermal motion—is a temperature that isn't present on Earth.

Everything above absolute zero, that is, everything on this planet, puts off heat energy. And everything that puts off heat emits light of some kind. It is just that, in some forms, it's via wavelengths humans cannot see, because we do not have the sensitivity to do so, though other species, like pit vipers, have organs that allow them to read longer wavelengths. Night-vision goggles mimic this for humans, amplifying visible light, giving us a peek at how pit vipers might view things.

Richard points to a cinder block sitting near his porch, imbued with the heat of the day, as something emitting infrared light right beside me. Though we cannot see the waves it is emitting, if we touched it, we would still be able to sense infrared light as heat. "Even if the sun went out, there'd still be infrared radiation around," he tells me.

The temperature of the universe at large is generally considered 2.7 kelvin. That is above absolute zero. This means that even what we view as the black part of the night sky is not an absence of energy. It is, arguably, a state of calm in an increasingly frenetic and artificially lit universe.

But what is darkness, really? And how have I spent nearly half a century not knowing? "If light is energy traveling at us and darkness is reduced energy, does that mean that darkness is a form of rest traveling at us?" I ask.

"I suppose you could think of it that way," Richard says, smiling slightly at the notion of night as a care package delivered on the regular.

What we're talking about evokes the tenets of animistic religions, nature-based traditions purporting that spirits reside in all that exists—in every stream, mountain, and tree. Once, in Japan, I visited a Shinto shrine where I was advised that there were spirits, kami, present in local geology. And now I know that all the while, mountain stones had been emitting unseen light all around me.

Light that we cannot see is emitted by everything that we can see. The good. The bad. The ugly. The things we consider inanimate. The things we think of as being dead. Energy is present even in the black space around stars, just as firefly larvae are present in my backyard, glowing brightly in the ground even during the coldest nights of winter. Even in what we call true darkness, the earthly places where our eyes find rest, light energy does not disappear. It simply hushes.

"We have the ability to see this tiny part of the electromagnetic spectrum," he says. "Because that's all we can see, we tend to think that's all there is. But there's so much more."

All the Life We Cannot See

Barely a week passes before I'm cradling the remnants of the screech owls' nest in my hands. First, the neighborhood owl box was hit by artificial light. Now, in a twist that's hard to believe, the owls' tree has been hit by lightning.

The bolt came just as the last fledgling was strong enough to fly out on its own, so no owls were harmed. The Coopers were out of town when the surrounding hills echoed with the sound. The oak was split into two halves, near perfect in division. The hardwood nest was cracked like an egg.

I've been dispatched to inspect the damage. The owls' home is a complete loss, but the human houses around it are intact. After reporting to anxious friends, I started picking through leaf litter to discover the bits and pieces of what was once the birds' sole domain.

The nest's interior is an exposed, previously dark place that I wish I'd never had the opportunity to peer into. But here it is. And here I am, attempting to decipher its contents like a fortune-teller. I find bits of owl pellets and hollow-quilled feathers embedded in wood grain.

In the shade of trees spared, I recall that Athena, goddess of owls, daughter of Zeus, was among the few Greek deities said to be able to cast lightning from her fingertips. Sarah, who wears an ancient-world Athenian owl ring, now has an owl nest smote by lightning lying inches away from her patio table. The chances are unfathomable. The splintered wood, difficult to bear.

In ancient times, lightning—like the appearance of an owl—was viewed as a message from another realm. Maybe it still should be. It's estimated that lightning rates will increase 12 percent for every degree of rise in global temperature in the future. That means by the end of the century, Earth might have experienced a 50 percent increase in lightning strikes since 2014. This will bring wildfires, since the U.S. Forest Service reports that 45 percent of them are lightning-started. Forest fires can themselves create lightning when they get going, due to particulates of air pollution rubbing together, charged. The larger the blaze, the harder it becomes to extinguish. The more lights we turn on, the more there are to turn off. On and on.

Artificial light at night even impacts daytime air quality. In natural night, compounds called nitrate radicals scrub volatile organic compounds from the air before they have time to turn to smog. It's a process that cannot occur during the day, since sunlight destroys required molecules—as does artificial light. Over Los Angeles, artificial light at night is now so bright that levels of the natural nocturnal cleanser have been reduced by up to

4 percent. Basically, under the influence of artificial light, atmospheric molecules are unable to clean the sky in preparation for the following day. They can no longer preen.

The atmosphere that has protected us by holding out the most harmful wavelengths of sunlight is now visible in night photography as holding in artificial light. From the right angle, light pollution can be seen as a hard-shell coating on what was once the periodically dark-facing side of our planet. From certain vantage points, Earth now looks like it's being cooked from the inside out.

When I first heard the story of Rocky, the saw-whet owl who escaped the LED-lit Christmas tree to head to Appalachia, I was romanced by the notion of sky islands as remnant habitat from a long-ago climate shift. But in this world of increasing light, in this time of climate change, maybe all patches of darkness are remnant habitat, a term generally used to refer to wildlands that have not been severely degraded by human activity, even as the areas around them have deteriorated.

Undoubtedly, somewhere close to where I'm crouching, the screech owl family—which I watched grow over time—is now roosting in shade. Sadly, there is no way to quickly grow another oak tree with a nesting hole, gift of an aging tree. But, now that the streetlight has been removed by the town, the dark habitat they lost to light at the start of this season has been restored. There is solace in knowing that these owls might have a safe place to land next season—in relative darkness, with an owl box waiting. The residents of this neighborhood have considered both nesting locales and dark-sky habitat. And it's likely that, next season, their efforts will be rewarded with serenading.

Around the world, people are working to bring attention to

the importance of natural night. In 1988, a pair of concerned astronomers founded the International Dark-Sky Association—which was later renamed DarkSky International—to "preserve and protect the nighttime environment and our heritage of dark skies through quality outdoor lighting." It was the first organization to mark the dark-sky movement, which is a global campaign to reduce light pollution.

One of the organization's best-known programs is accrediting International Dark Sky Places, an initiative founded in 2001, to encourage communities to preserve darkness via responsible lighting and public outreach. Through the program, towns and cities can become "International Dark Sky Communities," and there are other labels for wilderness areas. This language evokes the National Park system, with places set aside for darkness to reside. But, not unlike the saw-whet owls who use tree farms to hop from habitat to habitat, dark-dependent creatures need connected corridors to migrate, just as they need pools of darkness in which to nest.

In Europe, there's a growing movement calling for "dark infrastructure." Advocates suggest that, in the way green infrastructure focuses on land conservation and blue infrastructure on the health of waterways, dark infrastructure should focus on nocturnal habitat. This includes finding ways to map light pollution and preserving and restoring darkness with an attitude of temperance that some have started referring to as "lighting sobriety."

Birds of various species have been found to nest up to a month earlier in areas of light pollution than those living in places of relative darkness. This can be disastrous for some, since the insects they depend on for food are not available, leading to starvation.

For other species, it is a benefit, because the insects they depend on are also aligning with faux daylight rather than the cues of darkness. We're all increasingly following the directives of artificial light rather than the messages of the universe. Because often, in the human-generated din, we can no longer clearly perceive them.

Dark space is important not only as reservoirs safeguarding nests but as passages of safe travel. Winged creatures need dark skies the same as migrating paws and hoofs need unfenced terrain to tread. We might think of songbirds as daytime companions as opposed to night owls, but 80 percent of North America's migratory birds travel at night, using stars as navigational devices while avoiding the turbulence of daytime thermals. In U.S. cities alone, 365 to 988 million birds are killed every year during these nocturnal migrations in part due to artificial lighting issues, which disorient and cause them to collide with buildings, often fatally. And without access to navigational stars in cities awash with LEDs, some birds simply lose their way, with city lights drawing them away from their ancestral migratory flyways.

Ornithologists employ various methods to track bird migrations, including tagging birds, using geolocators, and satellite tracking. One way to detect birds' presence is by weather radar, which can help scientists identify peak activity so that they might warn municipalities to turn off lights on particularly active nights. In this way, via computers, scientists probe bird behavior by studying the presence of animals as data translated into light. But not always.

Even now, in the satellite age, ornithologists still employ old-school fieldwork that shuns artificial light for methods that immerse them in night. Via a practice called moonwatching, they sometimes fix scopes on the surface of a full moon. Through the scopes, birds can be seen as passing shadows.

This practice was popularized in the late 1800s after an ornithologist happened to be given a tour of a university astronomy department during an active migration night. When he was offered a close-up view of the moon, he was shocked to see migratory birds slipping across its face. Inspired, he started figuring out ways to estimate flying altitude, using moonlight to help him solve earth-based mysteries. The number of birds observed in a moonwatching session gives an idea of larger migration activity.

Around the time that moonwatching gained popularity, in the 1940s, nocturnal flight calls were recognized in migration monitoring. Contemporary moonwatching ornithologists often carry field microphones in buckets that act as amplifiers so they might discern which species are flying above. By catching songs in quart containers, the way that, as a child, I tried to capture slippery minnows, birders can discern the specifics of what they're witnessing.

Before I spent time with owls, I had no idea that there are millions of animals, rivers of life, rushing through darkness, clear across North America, every spring and fall. Some birders claim nocturnal migration as their favorite time of year, since they never know which species they'll find in their own yards as creatures pass through, off to nest elsewhere. Those travelers often belong to far-off flora, but they need dark skies as through lines.

A few hours from now, just down the road from where I'm crouched in shade clutching the blackened remains of an owl nest, Richard will step out into his front yard to see if it might be a clear night for peering into the universe. Around the same time, moonwatching birders will be stepping into their yards, curious about migratory birds, since spring nesting and migration seasons coincide.

After Richard determines if he's going to send out his telescopes, he'll go back inside his house to probe the far reaches of space via his computer. Some of the ornithologists might return to their computers, too, to look for birds in the form of burning clouds on weather-radar screens. But those employing the techniques of moonwatching will necessarily stay outside, away from artificial light.

An untold number of migratory birds will be moving over this neighborhood tonight, and the natural darkness of this street stands to protect them. They'll be carrying birdsong to far-off places, to people who have no way of knowing that the chirps and whistles in their yards were, in part, made possible by this dark oasis. They might not even realize, as I didn't until recently, that many of the brilliant feathers they appreciate in daylight were delivered by the winds of evening.

Glowworms Squirming

Local Tourist

I've got maggots on my mind. This might not sound pleasant. In fact, it might evoke a reaction along the lines of: *Ewww!* But, if you've never met any fungus gnat larvae, that might just be bug bias talking.

Orfelia fultoni, more charismatically known as glowworms, are associated with damp, dismal places, because they require darkness to shine—which they do, a deep, brilliant blue. On cool spring evenings, these maggots make earthen embankments gleam as though they're encrusted with sapphires. I've seen them once—and once wasn't enough.

It was in Tennessee that I first encountered the species. I was walking a path along Norton Creek in search of blue ghost fireflies when I saw light glimmering on a raw road embankment. In awe, I told Will Kuhn, the biologist who identified them for me, that I'd always wanted to visit the famed glowworm caves of New Zealand.

Because of this, it was particularly shocking when Will told me that Southern Appalachia is one of the only other places on Earth where blue glowworms are known to exist—and that the *Orfelia fultoni* I'd found on my firefly quest were cousins to the famed glowworms that live in New Zealand and, to a lesser degree, Australia.

Will admitted that he, like me, was guilty of not always recognizing the wealth of life around him. Being cognizant of this, as the head of Great Smoky Mountains National Park's biodiversity inventory, is his actual day job, but it wasn't until the pandemic that he'd started recording species in his backyard after his farther-afield projects shut down. "It's just a regular-size yard," he told me. But already he's identified hundreds of species that he hadn't noticed before.

If I was surprised when Will told me that there are populations of glowworms across Appalachia, I was flabbergasted to later discover that, in the mountains of Alabama, some reside in wondrous sandstone slot canyons. Alabama's Dismals Canyon holds one of the largest known populations of *Orfelia fultoni* in the world, with glowworms present in spring and summer. The species is so closely associated with the area, people occasionally refer to the glowworms as "dismalities," though the attraction is not well-known outside of the Southeast. Still, *Orfelia fultoni* spin webs on slot-canyon walls, without fanfare, embraced by darkness. And I aimed to find them there.

Only, when I started travel planning, I learned that the season had been cut short by a heat wave. Where once there would have been thousands of glowworms, there were only a few still shining. Locals feared that, in a matter of days, numbers would dip below a dozen. This made searching for glowworms feel like wonder-chasing during an apocalypse. The indication of climate change was harrowing. The personal missed opportunity, beyond disappointing.

Then, almost exactly a year from when I gained my first glimpse of *Orfelia fultoni*, Grandfather Mountain, the local-to-me place where synchronous fireflies were discovered in 2019, announced that they were hosting an inaugural event called Grandfather Glows to celebrate bioluminescent creatures. In advertisements, glowworms were noted, in fine print, as a sideshow.

Event organizers knew people would be interested in attending. But on the day that registration opened, so many people logged in from around the world that it crashed Grandfather Mountain's website. When it was relaunched, tickets sold out in 60 seconds. Despite the odds, with only a few hundred allocated spots, I secured one.

More than a few people I mentioned this to were envious, and one woman I know was downright upset, because she had tried and failed to get a ticket. She was mad that the mountain was charging money for nature communing, which she argued should be freely accessible to everyone. She told me the whole thing reeked of privilege. Then she revealed that, though she couldn't figure out how to find local maggots, she had previously visited the glowworm caves of New Zealand.

Despite her around-the-world encounter with their cousins, she hadn't even heard of *Orfelia fultoni* prior to Grandfather Glow's advertising. I told her that I've come to understand that diverse fireflies are present in pockets all over the region, suggesting that she might also have *Orfelia fultoni* along her rural road's embankment. It's a possibility I have been wondering about myself. Because night is unmapped territory. And, no matter where we live, no matter how otherwise outdoorsy and capable we might be, many of us are unsure of how to navigate even the land we know best when it turns into a nightscape.

～～～～～

I've been visiting Grandfather Mountain since I was a child. Always, I have related to the human-face profile of the mountain like the features of a beloved elder. After driving through Boone, I find peaks cresting clouds, making Grandfather look like a green-bearded man in a milk bath. I can see his nose, forehead, and green chin—each of them sky islands likely inhabited by saw-whet owls.

I turn onto a gravel road to enter the first-ever Grandfather Glows event. Almost immediately, I bump into John Caveny, Director of Education and Natural Resources for the Grandfather Mountain Stewardship Foundation. I've been acquainted with John for a few years. He's almost always sporting facial hair and a baseball cap, periodically claiming that he doesn't know a lot, but that he's good at finding people who do. Every time he says it, in a self-effacing way, I think about how this sounds, to me, like a leadership trait.

On a hillside where we can see visitors milling around educational booths, John tells me that during a recent staff pre-showing, it was a little windy for fireflies, but the glowworms were out in full force. "When everybody saw the glowworms, they couldn't stop asking questions. I think they had some idea of what fireflies might be like, but those glowworms were like nothing they'd seen before. Everyone kept asking me about larvae. When do you get an opportunity to talk about larvae? It was awesome! People are coming for the fireflies tonight," John says, "but I suspect they'll leave talking about the glowworms."

The species of glowworms found on Grandfather Mountain have two lanterns, and they are the brightest known blue-light-emitting life-form on Earth. John's seen them on riverbanks and roadsides, both locales bare of foliage. "We don't really know much about glowworms. Are they there because those are the places they prefer, or can you just not see them other places because plants are covering them?" he muses. "Have we just not found them in other places because there's no one looking? Nobody knows!" There are indications that the glowworms' bioluminescence is to attract prey, which then gets snared in the gossamer webs they spin. John isn't totally convinced that the light doesn't have other functions, but these glowworms are nearly unstudied in the context of their native habitat.

Fog begins to wrap around us like the graying innards of an antique quilt. If it rains too hard, fireflies might decide not to fly, but the glowworms will be delighted. John turns his face upward until he's a human-scale replica of Grandfather's profile. "It's a new moon tonight, pretty dark out," he says, "and these clouds should help us see some things." Glowworms, like salamanders, inspire appreciation for rain.

The thick mists of Grandfather might be a boon for glow-worms, but they make it hard to see Harvey Lemelin, one of John's expert contacts, until he's pretty much standing next to us, raincoat hood cinched around his face. He's traveled from Canada to attend the evening's event, but he's more concerned about the visitors who vied for tickets. "I hope this rain lets up, for their sake," he says.

Harvey, a professor at Lakehead University, is here as an

ecotourism consultant. He was particularly happy when John reached out to him about an evening event because he's observed a desperate need for after-hours engagement among students. Harvey used to talk to his classes about catching fireflies to evoke wonder, but every year, he finds fewer of his students have those memories to draw from. So he's had to replace bioluminescent encounters with examples of human-made light. "They've been to rave parties. So now, in class, I tell them to imagine glow sticks in the hands of people who are dancing. It's an image they're more familiar with," he says. "When we disconnect from nature, we disconnect from night. And remember, these are students who've chosen to take outdoor recreation classes. If they've lost that connection, it doesn't bode well for everybody else."

Even more concerning is that he's watched students, year after year, become increasingly filled with dread when he keeps them outside on evening field trips. "No one experiences massive darkness anymore. No one uses their own senses," he says. "That's where events like this come in. They give people an opportunity to reflect."

Harvey got his start researching polar bear tourism with a friend of his who was a biologist. "My friend was studying what the bears were doing," he says. "And I was studying what the humans were doing." He is the reason that polar bear tour vehicles have two guard grids in the open vent holes that allow tourists to view bears. When he saw operators' single-screen configuration, he suggested fortification. Otherwise, he warned, people's fingers would go all the way through, reaching bear jaws.

When he shared this double-grid suggestion, an operator asked if he thought tourists were idiots. "I told him, 'No, it's not about being idiots. It's about instinct,'" Harvey recalls. "'We're tactile. We instinctively want to reach out and touch things. That doesn't make us idiots; that makes us human.'"

These days, he specializes in entomotourism, a subset of wildlife tourism focused on insects. Harvey has written whole articles about dragonfly tourism, in which people specifically seek out the dragonfly larvae I found in ephemeral pools by happenstance. Those tourists are sometimes called dragon chasers, which gives the practice mystical flair. It doesn't take long for me to realize that he coauthored one of the firefly articles that inspired me to visit Great Smoky Mountains National Park to see synchronous fireflies in the first place. Entomotourism is a small field, and in it, Harvey's kind of a big deal.

With shrinking large-animal populations, entomotourism has turned to charismatic microfauna to foster more-than-human understanding. At its best, the practice moves people from bug disgust to tolerance to full-on insect respect. And, in the case of lightning bugs and glowworms, hopefully a new appreciation for natural darkness as undervalued habitat.

When Harvey was called in to consult on the planning for this event, he helped the mountain come up with strategies to mitigate the damage visitors might cause. "We could have put hundreds of additional people out here every night," John says. "But we're trying to figure out how to introduce this to as many people as possible without having them destroy what they've come to learn about."

Harvey gives a nod. "When all goes well with an event like this, we witness a transformation in people," he says. "They start to wonder: How can I make my own part of the world better? How can I bring this home?"

Beyond where we're standing, a trail has been sectioned off with rope. It's lit with red lanterns, making the scene feel like a film premiere. With darkness setting in, we follow visitors up the demarcated grassy path to a hardtop road, where some people have set up encampments, finding various ways to pass the time before sunset. There's a woman reading in a mini–living room she's built with a tarp. There's a couple walking, arm in arm, barefoot, down the blacktop road, splashing as they go. Harvey raises his eyebrows at the shoeless humans. "I'll have to make note of all this," he says.

We're not talking loudly, but the farther we walk, the quieter visitors get. Soon, we're shushed by a group of people who are already staring into the forest, looking for firefly lanterns, despite how much daylight is left. Harvey takes it as a sign that Grandfather's visitor guidelines are working. If the synchronous event at Great Smoky Mountains National Park was like waiting for a parade, the general mode of pilgrims here is more that of parishioners waiting for a sermon to begin.

Given the fact that I've been biding time with people in official gear, when I briefly separate from the Grandfather crew, a man approaches me, clearly thinking I'm part of the naturalist staff. He's a transplant from somewhere abroad where bioluminescent beetles are uncommon, and he's nervous. "Do the fireflies bite?" he asks.

At first, I think I've misheard him, but he repeats himself, arms crossed, finger to his temple as if, no matter what I say, he's determined to remember. "No, these fireflies don't bite," I assure him. "They just blink. They're not going to hurt you."

Relieved, the man leaves to share this information with a huddle of elderly women who are accompanying him. They chatter in whispers, some of them turning to smile at me in relief. I try to imagine how it might be to see fireflies without a childhood of spending time with them. There are surely things in this world that I unnecessarily fear, things that these visitors, familiar with other parts of this planet, know to be safe without hesitation. Out of context, it isn't all that difficult to imagine fireflies as little heat-seeking missiles ready to burn skin.

People often talk about fear of the dark as tracing back to ancestors, how threats of large animals at night probably made our species jumpy. But it seems far more likely that our way-back predecessors did not fear darkness as we do, because for them it was a place to which humans belonged. Our ancestors, all of them, knew the night as it existed directly around them, just as they knew how to identify the creatures and nuanced sounds that it held. How strange it is, from that perspective, to think that anyone would view darkness as something pulse-quickening in and of itself. Fireflies do not bite—but, then again, neither does night.

When I circle back to members of the roving event staff, I share the man's unexpected question. John doesn't seem surprised. In fact, the idea that a visitor has asked about a firefly's ferocity delights him. "That's exactly the sort of person we want to get out here," he says. "Someone who is afraid of fireflies is someone we want to be able to reach."

Harvey has fielded too many questions from undergraduates to be fazed by any inquiry. But, so closely juxtaposed to stories about his time with polar bears, I cannot help but think about how Grandfather is one of the only places I've personally seen a black bear free-roaming. Yet about those giant nocturnal creatures I have not heard a single person express anxiety. It's not that I'm particularly concerned about bears tonight—especially given the bustle of three hundred people, quiet though they might be—it's just strange to think that we've been so greatly distanced from darkness that we often do not have the knowledge we need to sort night terrors, much less wonders, by reasonable degrees.

After my neighborhood bobcat encounter, I did a little sleuthing about darkness and light as they relate to predatory behavior. In a study on bobcats, nights of low moonlight yielded less productive hunting than nights of high illumination. In the wild, increased light often helps large predators on the prowl, whereas darkness offers avenues of escape for prey. How strange that we tend to think the opposite.

Given that bears were not included in the promotional materials, I start wondering if visitors have even registered that bears live here. I'm inspired to ask Harvey, polar-bear consultant, dark-recreation educator: "Have you found that people are generally more afraid of bears or the dark?"

"Darkness," Harvey says, without pause. "We've almost completely forgotten how to relate to it, and we're alienated from pretty much all the animals that depend on it."

Harvey, unlike his young students, has caught plenty of fireflies in his day. But when it comes to connecting with night, it's always nice to find new, positive associations. He, as much as anyone, knows how fortunate we are to have found our way here. Harvey estimates that Appalachia's glowworm cousins account for one million annual tourists to Australia and New Zealand, where conservation concerns have led to limited viewing. It's one of the reasons John felt it was important to bring Harvey on board. Regionally, firefly tourism has already presented challenges, particularly south of here, in DuPont Forest, where public fervor to see blue ghosts led to forest-floor trampling—a sad reminder that a loss of habitat leads to a loss of magic.

Once a natural wonder is discovered, those who care for it often find themselves in a race to protect it in double-edged situations. Ecotourism can create acute dangers, but it stands to help lessen larger systemic dangers like light pollution. Studies clearly show that people are more likely to take pro-environmental action if they've had tangible encounters that make them care about a particular place, species, or issue. Even so, the balance of harm and reward is one that seemingly no one has tuned to perfection. But plenty of people are trying—and, on Grandfather Mountain, the level of care is heartening.

In addition to Alabama, Appalachian glowworms are known to live in Tennessee, Kentucky, and Virginia. But Grandfather Mountain is, as far as anyone knows, the only place in North America where it's possible to see synchronous fireflies, glowworms, and blue ghosts simultaneously. Harvey will, in official reports, refer to this evening as one of the first case studies of firefly-glowworm entomotourism in its "embryonic phase." I have not yet seen an iota of glow on Grandfather Mountain. But I am already witness to the birth of an international tourist attraction.

As a travel-loving human, I have done a thousand amazing things in a thousand different places around the world. I have, for much of my life, overshot in my quest for awe, overlooking the very land that gave me a love of place from the start: the mountains I explored as a child, not as sweeping vistas, but, rather, through the soft undersides of rhododendron leaves, the smooth-on-skin pebbles of creek beds. Now, people from all over the globe are overshooting their home habitats to discover the nocturnal marvels of my homeland alongside me.

~~~~~~~~~~

Grandfather Mountain's synchronous fireflies have been whoosh-blinking for some time before it's dark enough for me to see neon blue on a road embankment. Glowworms come into focus slowly until, finally, the roadside is a luminescent ribbon. A few stray blue ghosts are hovering above the stationary glowworms, creating some confusion. Are these two species somehow connected? Each time I've seen glowworms, they've been accompanied by these ghosts. It isn't clear if this is a coincidence or if there's some other relationship at play.

Some areas have more glowworms than others. I walk the road until I locate the largest group. Around each pinpoint of blue light, there are webs spun across soil. It makes the glowworms look like they are floating in cotton candy clouds. I get as close as I can to the road's white line to study them, navigating a gauntlet of chairs abandoned by other visitors.

The brightest area is partially blocked off by a family of four—mother, father, two teens. They were here before sunset, so I'm curious to know if they knew something I did not when choosing a location. I decline their offer to move chairs aside, but it emboldens me to ask: "Did you know that you were going to have the best seat for glowworms?"

"It was raining pretty hard when we came out," the dad tells me. "So we put our chairs here for shelter." I can, even at night, sense shade here, where tree overstory reaches all the way across the road, branches intertwined like hand-holding fingers.

These parents have flown in from the West Coast because,

as Tennessee natives, they have been missing the fireflies of their youth. "Grandfather Mountain was the place of so many childhood memories," the woman says, "when we saw this event, we decided to try to get tickets." But even though they've come for the synchronous show, they're sticking closely to the glowworms.

I tell them that, though I'm a huge fan of fireflies, I've mostly come for the glowworms. As we chat about bioluminescence, I mention that I've heard their home state has notable bioluminescent waters—full of organisms that light up when disturbed. "We'll have to look into that," the woman says, suddenly distracted by the arrival of a blue ghost. "As a child, I always wanted to know more about fireflies, so I'd trap them and put them in Mason jars to make lanterns," she says. "Back then, I didn't know their adult phase only lasted a matter of days. Days! I did things that meant they spent their entire lives imprisoned!" What was for her a blip in time was for that firefly a lifetime in captivity. "I think I even smashed some of them. It's horrifying when I think back to it," she says.

Currently, every firefly flash feels like something fragile and precious. So, too, does the dark cathedral of trees that's protecting us from the lights of housing developments beyond the gates of Grandfather Mountain. The woman says of her sons, who are hovering nearby, "When we got here, they were like: 'Are we really going to sit here for hours in the dark staring at bugs?' Now they don't seem ready to go. It's amazing." Their sustained interest is, to her, nearly as fascinating as the glow-world we're immersed in.

Since we're in such a populous glowworm spot, other visitors are drawn to it. "I knew about the fireflies, but these glowworms might be my favorite part!" a passerby says.

"They look like a night sky," another observes.

"I think it's more like being in an airplane, looking down at a city," their friend retorts.

Most visitors move on after quick encounters. But others linger. One woman, after crouching at close range, says, "Wait, are they moving?"

Her companion confirms: "I saw some shifting around. I thought something was wrong with my eyes at first!"

It is disorienting to have your gaze fixed on constellations or city lights only to have them shift. The worms, barely larger than the half-moon of my thumbnail, are writhing just enough to remind us that they are living light.

To me, they look like an almost exact match of Pleiades, a brilliant star cluster that shines 444 light-years away from Earth. These blue worms, nestled in their webs, and depictions of those blue stars, nestled in nebulae, are a perfect match in overlay. The resemblance is so uncanny that, forevermore, when I see space photos of Pleiades, I will instinctively think, at first glance, that I'm being presented with a photo of Appalachia's gorgeous, glowing maggots—stars that squirm on the surface of our home planet.

~~~~~~~~~

The Orfelia fultoni *species was* first recorded by science in the 1940s by B. B. Fulton, a naturalist who noticed blue light in his western North Carolina backyard and groped around in the dirt, investigating. He was the first person to document that the larvae conceal themselves in dark crevices during the day, inching out to lounge on gossamer webs at night.

Fulton attempted breeding experiments, but he quickly found

that, when exposed to heat, the creatures died. Around the world, glowworms that shine by night and require dark forests are, despite the pressures of climate change, still being discovered. It wasn't until 2020, a year after the discovery of synchronous fireflies on Grandfather, that glowworms were confirmed to also live here. They were initially noticed by my buddy John himself, which is a detail that he doesn't confess until several hours after dark.

He's taken a break from his rounds to join me under my tree shelter. I tell him that I've been thinking about how trees that offer shade by day also offer protection against light pollution at night. "I've never really thought about shade at night," he says. But, now that I've mentioned it, John recalls that he was in the layered shade of rhododendron, crawling around on the ground, when he found his first glowworm on Grandfather Mountain.

"At first," he says, "I thought I saw a female blue ghost waiting around." To be sure of what he'd found, he enlisted the help of Clyde Sorenson, the entomologist who had reported the mountain's synchronous fireflies. "It was pretty amazing when he confirmed that I'd found a glowworm, given there are so few places they're known," John says.

Globally, maggots seem to be having a moment. In 2019, the same year that I first saw *Orfelia fultoni*, a new glowworm species was described in Brazil. *Neoceroplatus betaryensis* is a confirmed cousin to both Appalachian and New Zealand glow-worms, another fungus gnat species, considered to be the first blue-light-emitting insect known in South America.

Orfelia fultoni's South American cousins' bioluminescent functions are thought to be identical. Brazilian researchers think the worms have potential uses in biotechnology, marking cells or genes, and in the creation of pollution biosensors. But, in a way, all glowworms are biosensors, since if you notice them, you know that you're in an area where light pollution has, at least in part, been held at bay.

Still, glowworms are such a fringe species that there still aren't many people paying attention to them. Even here on Grandfather Mountain, where things are being carefully considered, it isn't clear how many are in residence. "How many glowworms do you think there are here?" I ask.

"On this whole road? Thousands and thousands," John says. "We haven't done any surveys on glowworms yet, because we've been so focused on fireflies in preparation for this event." He makes a casual grid-frame with his hands and says, "We can safely say there are hundreds of glowworms here, just right in front of where we're standing." He drops into a squat. "But, Leigh Ann, check this out. Down here, things get complicated."

Standing, it's only possible to see one angle. But, if you're willing to shift perspective, there's more light to be found. I take a lower vantage point and gain access to glowworms that, from above, are shrouded by soil. I see new lights—more glowworms. Some are dwelling at the rims of the earthen holes they inhabit; others are more deeply embedded in their worm-size caves. Every time we move, more light comes into view. It isn't possible to see all the glowworms at the same time.

We zoom in and out. We go high, we go low. With worms, we wiggle. Soon, a couple of visitors approach, oblivious to our awkward posturing. As they study the glowworms, one of them says, "Hey, look, some of them are in little groups. They're like, 'Hey, we need friends too.'"

"Yeah, but then there are some out there on their own, like they're spreading."

John's filing away what he hears as things that might be worth looking into later. When the visitors ultimately move on, he says, "Just in this area, what I thought was fifty glowworms at first is more like a population of a hundred and seventy-five," he says. "I just counted. And that's in an area that's barely twenty-five feet wide. This quickly goes from hundreds to thousands when you take in the whole of this road." Upland, the road curves. We could not see the end of it from here even if we tried.

There could be, as indicated by a quick layperson's count, as many—maybe even more—glowworms here as there are in peak seasons at Alabama's Dismals Canyon, where, depending on weather conditions, populations range from 1,000 to 20,000. And this is the first night that the public has ever encountered them on Grandfather Mountain. Being here feels like participating in history.

Beyond the glowworm-encrusted embankment, synchronous fireflies are still making small whooshes of light. There are even a few blue ghosts drunkenly weaving in between the lanterns of other species. But John is, like me, wholly focused on the maggots that have turned this roadside into an effervescent river of light.

By day, this roadside is passed by hundreds of thousands of visitors a year en route to better-known attractions. After hours, it's frequented only by security officers tasked with protecting the mountain from off-hours trespassers. All night, the guards cruise around in patrol vehicles, headlights on high, blind to the life around them. At least, that's what they used to do.

When John first started doing firefly surveys to get a baseline population count under Harvey's guidance, he didn't have enough late-night helpers, so he enlisted members of the security team. "They were out here all night anyway, so I asked them to go to certain spots, get out of their cars, turn their headlights off, and report what they found." And what they found was diverse bioluminescence, all over the mountain.

Now the night watchmen of Grandfather Mountain have seasonal duties that go beyond paying attention to threats; it is part of their job to find and guard glowing lives. In sensitive areas, they are careful with where they even point flashlight beams. Because glowworms have helped them understand that, in preserving life, security lights are not always as important as the security of natural night.

The Cartography of Evening

Since entomologist Clyde Sorenson first reported synchronous fireflies on Grandfather Mountain, he has found yet another undescribed species—in his own lowland backyard. After a neighbor reported some strange glows, he decided to stake out the woodlands behind his house, where he's resided for more than 25 years. It was a site where he had, already, recorded at least 11 known-to-science firefly species. Yet when he looked again, he found fireflies resembling blue ghosts flying 200 yards from his own back door.

Are they blue ghosts outside of their mountainous range? Are they a different species? Are they new to science? He's started a citizen-science project to figure it out, admitting that the close-to-home discovery has been both delightful and embarrassing. He's a professional entomologist, after all, a renowned firefly enthusiast. If it could happen to him, it seems like it could happen to anyone.

I'm thinking about this one afternoon, not long after the Grandfather Glows event, as I walk the rhododendron-rich quarter mile it takes to get to my mailbox. The round trip to a main road takes some time, seemingly uphill both ways. Halfway through my walk, a neighbor slows his truck to ask if I need a ride. I decline, but I tell him about Grandfather Glows, and I explain that I've been scouting tree shade and trickling water and bare patches of dirt surrounded by hardwood leaf litter. After I detail my quest for glowworms, I ask, in all seriousness, "Do you think I should send out a neighborhood email saying something like: 'Hey, I'm going to be walking around after dark, please don't shoot me'?"

I'm on friendly terms with my full-time human neighbors, though our worldviews sometimes diverge. We alert each other to the timing of ephemeral wildflower blooms and host neighborhood parties with old-time bands. We often join each other on summer kayak trips, and in fall, we gather fruit from scattered apple trees to utilize the antique cider press that's stored in a barn every other day of the year.

We're not all that far from town, but far enough to know that depending on each other is sometimes the only way we can make it. We collectively endure blizzards and floods. But in darkness, out here, every animal is a shadow. No human is, from a distance, recognizable.

We must rely on local knowledge to navigate this terrain.

There are no Google Maps to help, no filters that can reveal hidden arsenals. I know which of my neighbors are stockpiling guns—and there are many of them. I know who locks them up when they're not in use, who uses a silencer when target shooting, who drinks at night and occasionally shoots at the sky for fun. I know who fled here from a city after being threatened with a machine gun. I know who has lived here all their life and recently taught their eight-year-old how to shoot deer, right off their porch, to fill their meat freezer for winter.

I don't have a count, but I suspect there are, in my neighborhood, more guns than people—all on a patch of land where houses and livestock and wildlife rub surprisingly close to each other. These factors comprise my reality. This likely sounds terrifying to people from off the mountain—and it is, admittedly, sometimes scary to me—but guns are part of an equation that I have, for all my life, been required to calculate when outdoor exploring.

During salamander season, I was nervous about visiting the gun club after hours, but in all actuality, in terms of raining bullets, I might have been safer there than in my own neighborhood. It is a place many people wouldn't want to tread after dark—especially if they didn't know it personally. This is, of course, another reason that Grandfather Glows is so sought after. Even when I visit the parks and trails that are considered public domain, I often notice closes-at-dusk signs. If night is disappearing due to artificial light, landscapes where it's permissible to tread after dark are also increasingly closed off by private property signs, security cameras, and, in many parts of the United States, gun-toting humans.

It might seem shocking, but—despite all the gun owners

around me—I still fear more deeply the harm of absentee real estate investors. There are some who buy property and develop it minimally, but many destroy flora and fauna and put fences up where—within my lifetime—locals were free to roam in an unspoken culture of communion. For generations, walking unposted forestland was a generally accepted practice in this area, so long as no one was harvesting things like ginseng or venison without permission.

My neighbor, who apparently does not overlay a mental map of gun ownership when taking mailbox walks, shrugs off my firearms concern. But when I tell him about my bobcat encounter, he gets nervous. "A bobcat, huh," he says. "Now, that's something you should worry about." In the back of my mind, I always carry a wise quote from my friend Mike: *The things we fear most aren't usually the things that wind up hurting us.* After all, we're generally afraid of the dark, but it's artificial light that might ultimately play a larger role in harming us all.

When my neighbor moves on, I go back to looking for places where tree canopies meet over the road. I scout ditches and drainage areas, slopes that have been left alone because contractors deemed them too steep to build on. This is a neighborhood dominated by gravel, paths that move underfoot to announce a traveler's presence.

In an area of relative shade, I notice some holes in bare ground. I see silken webs that lack the intricate patterns of spider work. I start attempting to conjure, in my layperson imagination, the cast of tiny creatures that might have crafted such a setting. Just when I decide this might be worth checking out at night, my dog, Wilder, an 80-pound goofball, bounds up a hill where hardwood leaf-litter meets road embankment.

I call him back with an urgency that he finds confusing, since—unlike the heart of my neighbor's vegetable garden—I've never emphasized this spot as worth safeguarding, but I don't want him to carelessly paw-churn the understory. In the wake of Grandfather Glows, this land—where spring water gathers in pools lined with mica dust—seems particularly precious.

~~~~~~~~~~

I am the kind of person who jumps at her own shadow. I know this for sure because, when I head out at dusk—accompanied by my reluctant twelve-year-old son, Archer—I leap at the sight of my profile, silhouetted by a neighbor's security light. Archer tells me that he didn't know jump scares existed outside of horror movies. His smirk makes me think that my antics have, if nothing else, made our search more entertaining for him. "What did you think that even was?" he asks.

"I thought I saw an animal," I say. "I mean, I know I'm an animal, but I didn't recognize myself for a minute. I guess I mistook my shadow for a bobcat coming after me."

He laughs, but I can feel him tense at the mention. This is a gravel patch where he himself has seen a fox trotting recently, and there's no telling who's around this evening. Researchers at UCLA surveyed 62 species across six continents and found that mammals around the world are fleeing daylight to become more nocturnal as they attempt to get away from human disturbance. It isn't clear if artificial light is creating the spatial-temporal changes or if the animals are simply desperate for some alone time with the human world closing in. In a rural community of increasing development, I can relate. Once, I met a new neighbor who recognized me from images he'd already captured on his wildlife camera.

I'm not sure if there's another mammal within earshot on the road, but Archer and I chat at normal volume. If there are bobcats or bears here, our voices, announcing human presence, will likely be enough to ward them off. And if any of our neighbors are out, they stand to recognize the sound of our voices as part of this landscape's familiar song.

We have already walked deeper into darkness without a light than I've ever traveled by myself, though we still haven't gone very far. I have rambled the globe solo, but I apparently need a preteen companion to feel emboldened enough to explore the far reaches of my own neighborhood after dark. My son is barely on the cusp of puberty, but he is already taller than me. And he is growing, growing. But no matter how much he grows—or how much I end up shrinking in old age—together we will always be bigger than we are alone.

There is a pole-mounted light below the cliff-side road we're walking. Due to our elevation, I'm standing directly next to the bulb. It belongs to vacation-home owners from off the mountain who visit, maybe, twice a year. I haven't met them. Neither have my other neighbors. But we know well the light they leave on when they're away.

Some spots on this mountainside are shaded by trees, while others are being hit by artificial light. I can tell that, precisely where the artificial light shines through, unshaded ferns have yellowed, likely doubly exposed to light during the day. These plants are never able to fully escape blue light. The vacation-home owners rarely use this place as a getaway. But even in their absence, there's no way for these mountains to take a break from their ecological impact.

I have started seeing new lights across the river, revealing houses I otherwise would not have known were there, because they're hidden by trees. It is as though people think of the lights themselves as protective. But where those dwellings are located, there's no one to witness what's happening. As it is, those security lights only stand to draw attention to what doesn't fit. If anything, they signal to locals that a house is standing empty.

I think of Harvey's beloved polar bear populations, dwindling. I think of the glowworms of Dismals Canyon, sizzled out of season. When we leave lights unchecked, we torture ourselves and everything around us, strangely unable to acknowledge connections between the burning of the world and our burning of midnight oil. Whole ecosystems are disintegrating in our backyards while we're worrying about the melting of far-off glaciers. It's harrowing. Still, Archer and I are out here looking for signs of hope. We're here to figure out what stands to be lost.

We slow our pace once we've passed the worst of the vacation cabin's security beams. At first, I think the sensation of cooler air is psychological, but I soon realize that we've entered a microclimate by simply walking a few feet. It's a quirk of mountainous terrain. This is an area of night shade.

Even in urban areas, neighborhoods with substantial tree cover can be cooler than others by upward of 20 degrees, though in some cities, sparse trees can no longer keep up with the need for shade, so activists have had to turn their efforts toward installing overhangs and umbrellas to create pockets of daytime darkness. But the effect isn't the same, because the cooling power of trees isn't just about blocking light; it's about how trees share their stores of moisture. Shade trees don't just offer a dry coolness; they create mist in a process called transpiration. The collective moisture and emitted isoprene that hover over Appalachian forests give these mountains their long-range haze. It's how the Blue Ridge and Smoky Mountains got their names.

"What's that?" Archer asks about a stand of flowers. Dark foliage has melted into night, leaving only white-dot petals.

"Elderberry," I say after a few minutes of study. Here we're both children, relearning. Some lovely things that blended in daylight now stand out, reflecting available light.

When we reach a crossroads, Archer thinks he sees a shooting star. I think it might have been a high-flying firefly. Ultimately, we decide that, for our purposes, the specifics don't matter. By either name, it was light in motion. Energy, animated.

I suggest that he should make a star-wish. It's a beautiful tradition, this act of sharing dreams with the sky. It's a practice that introduces us to the notion that we might talk to the cosmos directly, calling on powers that can move tides and seasons in hopes that they might intervene in our small human happenings.

In the woods, something tumbles. Branches crack. Leaves are hit. There's a squeak. Then, a squawk. "What was that?" Archer asks.

"It sounded like a nest of flying squirrels falling from a tree, or baby birds, maybe?"

"I think we should go," Archer says. But it is not yet dark enough to see bioluminescence.

Then a whinny from the same direction as the squeaks. "Did you hear that?" Archer asks, shocked by the amount of action in darkness.

This sound, I can identify for certain. This, I can teach.

"It's a screech owl. Listen," I say. And he does, rapt for longer than I expect.

Archer is hurrying out of childhood, and he's been cold-dropped into middle school after a confusing pandemic period. Some people might have fared well with emergency homeschool and virtual learning. We—neither of whom are gifted at creating the hard lines of structure, both preferring organic shapes and fuzzy edges—did not. He is a wide-open human in a world that doesn't often have time to remind him to breathe. And, frankly, he gets that from me.

I have the sudden urge to grab his hand, now larger than mine, as if he was still a young child and we were preparing to cross a street. But we are not crossing, we're purposefully standing in the middle of a road in the middle of the night, so I lay a hand on his shoulder instead. He does not turn away. If anything, he leans into it. A week ago, it would have seemed odd to imagine that we'd be outside, staring into darkness. Now, it seems weird that we haven't done it before.

Beyond us, the screech owl keeps singing to let family members know that they're surrounded. I give Archer's shoulder a squeeze. We're seldom given formal introductions to darkness as a locale, a place of nonthreatening presence rather than a store of fear or symbol of life's absence. From an early age, we are almost always taught to cower from it. But not tonight. Not here. I think, though I do not say: *This is darkness, part of your home, part of you. Do not be afraid.*

Darkness deepens. Archer reciprocates my gesture with his free arm around my shoulders. It's a show of preteen affection, and I cannot help but think the privacy of night has freed him to make it. We're clustered together, tightly as glowworms. My mothering heart feels, for a minute, brighter. But parenting is often about pivoting at short notice.

Soon, Archer is hungry. He's complaining about being tired. He is the sort of kid who dreams of moving to a city as soon as possible, and he has already put in a good deal of time inspecting dirt with me. But I ask if he can hold out for a few more minutes. Soon after, two blue ghosts appear on the hillside above us, bobbing and weaving. I cannot believe it.

He seems impressed with the ghosts' steady light, which is evocative of neon signs. "How rare are these?" he asks. "What would happen if I caught one?"

It is, after his years of summer nights spent catching and releasing common fireflies, hard to resist the impulse to reach out to capture these blue ghosts. It's hard to go against instinct.

I tell him about how synchronous fireflies were once thought to live only in pinpoint-specific places, but that understanding of their population range is growing. How, with blue ghosts, the same is happening. "I'm not sure if they're regionally rare, but they seem rare here. I mean, I'm only seeing two or three. It probably wouldn't hurt if you were careful. But let's just leave these."

Archer accepts my directive. But he is still hungry. He is still tired. Despite this, he takes a second to examine the road embankment more closely. My instincts have, in terms of living blue light, proven to be surprisingly on target. But, despite our concentrated attention, nothing. "Let's go to bed," he says. "We'll come back tomorrow."

We turn in the direction of our house. And it is in that field-of-view shift that I see them—tiny lights, nearly full-shrouded by foliage. I emit a high-pitched squeal. The blue ghosts have been a nice surprise, but I cannot believe that there have been glowworms shining among us, wholly unnoticed. I point them out to Archer. "You sounded like a teakettle when you saw those," he says.

There are no painted road lines to rein us in, only the threat of poison ivy. I turn on a red-tone headlamp to peer into the worm-size caves of my neighborhood, their tiny openings knit over with webbing. These neighborhood maggots aren't embedded in dirt. They're nestled in moss beds. "What do they look like up close?" Archer asks. I don't know how to respond. I don't have a good read on them, even though, by now, I've seen thousands from a distance as stars that elude close examination. In the macro photos I've seen, glowworms have translucent bodies with brilliant blue light emitting from both their heads and tails. Their glow cannot be properly photographed, in the way that sunset photos always make the phenomenon seem small, every two-dimensional attempt sad in comparison.

I want to pick this glowworm up. I want to examine it, show it to my awaiting son. I am drawn to claw at its cool illumination with the heat of my curiosity, but I keep my hands to myself. "I'm afraid I'll hurt it if I pick it up," I say.

"Then don't," he says. "I can see it from here."

Wishing I had more to offer, I say, "In the close-up pictures I've seen, they look like glow sticks."

I'd pitied Harvey's students who had never seen fireflies. Now here we are, realizing that we have only just been introduced to the glowworms that glow sticks resemble. In glowsticks—omnipresent at birthday parties and music festivals—my son and I have been accepting stand-ins for real things without knowing what we've been missing.

As we turn to finally walk home, we realize that we could not reliably find our way back to this exact location. Archer fumbles to find large pieces of gravel, which he stacks into a miniature cairn. "This will help us tomorrow," he says. "How long have you been trying to find glowworms, anyway?"

"Well, I've been thinking about them for a while." He doesn't comment, but I notice that he's walking with a little swagger. He was, after all, the first to spot the blue ghosts earlier.

When we reach the front door of our house, he releases a sigh that's reminiscent of my teakettle squeal. "*Ahh*," he says. "It's good to be home." It feels like we've just returned from a transcontinental voyage, though, in measurable distance, we've hardly traveled at all.

~~~~~~~~~~

When we locate the cairn the next day, in sunlight, we find that the place that we attempted to mark no longer exists. Archer's stone-stack, which appeared as an ethereal altar in the dark, now looks like a pile of dirty rocks. The palette of soil previously holding neon blue signs of life seems a dismal wasteland. And yet. As long as this area remains undisturbed, as long as trees are left intact and my absentee neighbor's security light does not burn surrounding foliage to death—we wouldn't be required to take a single step if we wanted to return to that wonderland. It is, already, spinning its way back to us.

I hover over a rock face that's emanating the stored cool of night. It feels like peppermint against my skin. I keep my hands to myself, but my gaze moves from dot to dot, connecting dewdrops caught in glowworm webs. I study the markings of a still-standing dead tree, likely home to a screech owl family. Without context and an earnest plea for taking care, it seems imprudent to leave a directional marker here. Stone by stone, as we dismantle the cairn, I suggest that Archer should join me in taking note of the fern species, the dripping of springwater. These are not clues that I've seen in any book; they're just my own time-and-again observations.

Grandfather Mountain's glowworms have been known for a while, but the ones in my neighborhood were, I'm confident in saying, first spotted in living memory yesterday. Because of this, it seems wrong not to share this place with the woman I know who was upset about not securing Grandfather Glows tickets. The place we've marked is a spot I know she has driven past a hundred times herself. And it looks very much like her mountainside driveway in a neighboring county.

Destiny might be written in the stars, but it's translated by geography. And it will be a special thrill to share with my frustrated acquaintance that what she sought an admission ticket for is a wondrous world in which she is already living. Because she, like me, is privileged to live in a landscape full of shady, mucked-up places. It is harder to find wonder close to home than it is in far-flung places; where everything is new, nothing can be taken for granted. But when exploring darkness, it's easier to be awed, wherever we stand. Because night is a foreign country, and I, for one, am seeking dual citizenship.

It has been well documented that we suffer, in all kinds of ways, when we're separated from the natural world at large. But the losses of night are, in the human story, ones we've only just begun to tally. We attempt to measure them in light exposure rather than darkness deficit, maybe because we think of darkness as an absence, nothing to lose. But darkness is, arguably, the genesis of life itself.

I'm starting to suspect that the anger my ticketless acquaintance felt toward Grandfather Glows—an emotion shared by many who did not secure entry—might have been less about that singular event than about the broadscale frustration of living in a culture bent on alienating us from the land we live on and that, on some subconscious level, we're all feeling the loss of darkness itself as a commons. There was a time, not so long ago, when humans could touch darkness without even trying. As recently as when I, myself, was a child, night was an altogether different place than it is for Archer.

None of us alone can reclaim natural night. But maybe reclamation begins by attempting to honor the lives of the tiny, sometimes shining creatures who depend on night deepening not just above us but also around us. For healthy ecosystems, darkness is required. Even in the land of the midnight sun, the opposite phenomenon of polar night arrives for balance—though those extended periods of darkness are lesser talked about. In my neck of the woods, glowworms stand to help us recalibrate how we think about dismal darkness, because it is the only place where their brightness shines. Maybe we can revitalize massive darkness by fostering the conditions required to locate our own homeland's minuscule wonders, wherever they might reside.

Leaving no trace of our amazing find, we head home. The land that holds our neighborhood glowworms is in a legal right-of-way, but the property does not belong to us on paper. Its deed is held by people who do not live within a hundred miles of this land, absentee investors who could likely not dream that they control the fate of glowing creatures that tens of thousands of people annually vie for tickets to witness, living marvels coveted by researchers as far away as South America.

Right now, day blind, even Archer and I no longer know exactly where the glowworms reside, but we don't mind. We no longer need a cairn or a hard-to-get ticket, because learning to read habitat has given us invisible-ink maps to entire kingdoms. We're only beginners, but we understand that routes drawn by shade and stone and water and moss can, in darkness, lead to treasures hidden.

Summer

Moths Transforming

~~~~~~

Shadowbox

Archer has found a shrunken, radioactive lobster in our backyard river. At least, that's what this crawdad looks like in the beam of a UV flashlight. We've seen plenty of crawdads in our day. But since they are nocturnal and it's well past ten o'clock, the river is crawling with numbers we didn't expect. And under ultra-violet, or UV, light—purple wavelengths that are even shorter and denser than blue ones—this animal seems to be glowing.

"Should I pick him up?" Archer asks. Before I can answer, his hand is in the water and the crawdad is pinching. Archer says, "He's really trying to get me!"

Archer doesn't hold the animal for long. Catching and releasing crawdads and salamanders and fireflies is how he's come to have a relationship with the natural world, though as he has gotten older, his time outdoors has lessened. It seems a sad passage, the notion that communing is something he will outgrow. Then again, I've been around for a long time, and I'm the one who instigated this wacky UV-light expedition.

Back in the 1990s, I had a black light that amped up the wow factor of posters that lined the walls of my bedroom, technicolor dreams hovering over incense burners. Because of this, black lights always rouse nostalgia. They were some of the first tools I ever had to gain perspective on how light spectra might alter the world—and the way I felt in it. So when I came across an article about biofluorescence, I was intrigued.

Bioluminescence occurs when an animal produces light through chemical reactions in its body. Biofluorescence occurs when an animal absorbs and re-emits light. It's found in reptiles, fish, birds, and invertebrates—and, to the recent surprise of many field researchers, mammals—as they absorb short-wave light and reveal it via skin and fur as longer wavelengths that are visible to human eyes with the assistance of UV devices.

The accessibility of UV flashlights has led to an onslaught of biofluorescent discoveries in surprising places, including New Jersey parking lots. I'm not sure if the crawdad species Archer has found is known to be biofluorescent, but it's certainly interesting. These flashlights have turned our backyard into an immersive hallucination. Out here, he's about as swashbuckling as a twelve-year-old can get.

When we step out of the river, our feet sink into loam, and we're swarmed by beating moth wings. I turn my flashlight off to stop the bombardment. "They must like UV light," Archer says. "I think they're attracted to it."

From behind a riverside buckeye tree, an animal starts clicking. I think it's a sound that racoon mothers make to calm their children when they're nervous. It's making me feel like I should tell my child to clear the area. But before I can speak, Archer's on the move.

"We need to get out of here," he says. It isn't the racoon that's spooked him; it's the insects. He's wearing a white shirt and they're flocking to it. One flies down his neck, becoming trapped between skin and cotton. "Better to be shirtless than to have that happening," he says, stripping until he's half-naked in a cloud of wings.

The hill leading back to our house is a steep one. As we climb, more moths appear, too fast for us to really see them as they bang against a flashlight. "I can't believe how much is going on out here," Archer says. The intensity of life in darkness, so often thought of as negative space, is astounding. Even after my encounters with fireflies and salamanders, there's a bounty of activity here that I, too, find somehow surprising.

Archer has spent the day at a county-run day camp. "I found a butterfly today," he says, matching my stride as we heave-ho our way uphill. "I mean, I thought it was a butterfly. But now I'm wondering if it could have been a moth. I've never really thought about moths before."

Twenty minutes outdoors with a UV flashlight and, already, his understanding of the world is being reshaped. "What did it look like?" I ask.

"It was big. Like, the biggest butterfly I've ever seen."

Once we're inside, steeping in artificial light, I pull up a photo of a cecropia moth, the largest moth species in North America. Archer taps my phone screen, enlarging the image of crescent moon–patterned wings. "That looks like the one I found!" he says.

"Amazing! Dead or alive?"

"Dead," he says. "But that moth was so cool I put him behind some rocks so the other kids wouldn't step on him."

"Do you think the moth's still there?"

"Yeah," Archer says. "I hid him really well. If I hadn't, he would've gotten smashed."

In recent years, Archer has evolved from a child who prefers to go barefoot to one who covets designer sneakers, but he, like me, still considers natural findings to be gifts. In our house, windowsills are crowded with artifacts that we use to document walks and river trips: a hummingbird's fallen nest. A robin's egg. Driftwood, sun bleached.

It's getting late. Still, this seems worth exploring.

"Should we go into town and try to find that moth?"

Archer tilts his head, curious. "Let's do it!" he says.

We've left our UV lights behind, but we're both wearing headlamps when we arrive at the recreation center. Archer pulls his light off immediately. "Guess we don't need these," he says.

It's nearing midnight. There's no one in the parking lot. The tennis courts are clear, but stadium-size lights are blaring. Under newly installed LED security lights, it doesn't feel like we're on a nocturnal outdoor adventure at all. It feels like we've driven into a surgical suite the size of an airplane hangar. It suddenly strikes me as strange that, given all the things that our culture claims people should be personally responsible for securing—health care, for instance—we're almost uniformly obsessed with providing communal lighting.

I suspect the bubble of light we've entered, in a parking lot as large as a downtown block, has something to do with the fallen cecropia moth. No one knows for sure why moths are attracted to artificial light, but one theory is that they use the moon to navigate, holding it as a fixed point of reference. Lightbulbs are concentrated energy that, mimicking the moon, rattle their senses.

Some researchers think that, without access to the true moon to guide them, they're at a loss about which way to travel. What's known for sure is that, when enticed by artificial light, moths don't know what to do with themselves. They fly in circles trying to figure out where they should go, how they might navigate. A moth that becomes disoriented by light stands to be, in essence, energized to death.

In this age of entertainment and never-ending social media scrolling, this feels a little close. For both moth and human physiology, artificial light is a stimulant. And, under the influence of it, I get caught in echo chambers myself. I'm not sure that I know anyone who doesn't, at least on occasion.

Archer leads me around the gym, which has a wall of glass lit from the inside with no blinds. The indoor light is met by the illumination of an outdoor security light. Everything in sight appears to be burning, and we are the only two humans here to see it. The grassy space where children play in daylight is here, in the depths of night, brighter than my living room with every lamp turned on. It seems, in fact, brighter than any place I've been in recent memory.

Pressed between the exterior of that plate glass window and the lamppost, there's a small swath of landscaping stone. And along its edge, sheltered by my son, the finder, is an intact cecropia moth.

This is a perfect specimen of the largest moth in North America, laid flat against stones that match its sheen. The moth, perfectly preserved, appears to be part of an open-air museum exhibit. Archer has carried a piece of paper from the car. He jimmies his fingers under the moth's wings and lifts the creature onto the paper, lest we damage wings in transport. On the drive home, he holds the paper with both hands open, like a wedding-ring bearer who's been entrusted with platinum bands.

To make it to this stage of life, the cecropia began as an egg last year, likely on the leaf of a maple or cherry tree, the plants they prefer. When hatched, a tiny caterpillar depended on leaves for food. In the span of a month, the caterpillar shed skin four times, in stages known as instars. In full-size caterpillar form, at the time of pupation, the caterpillar likely chose a branch to attach to, spinning a cocoon of silk that blended perfectly with tree bark. In this warm interior, the insect became a pupa. For ten months, this creature rested in that dark space, camouflaged from hungry animals. Then, almost a year from when the process began, the cecropia pushed out of a container of isolation to become what we're seeing.

A cecropia's adult life is short; their sole responsibility is to mate and ensure future generations. I find myself hoping that this moth connected with others before getting lost in municipal lighting. It's probable, since cecropia pheromones are detectable up to a mile away. This magnificent moth has, in all its forms, escaped ravenous squirrels and songbirds and owls only to be taken down by lightbulbs.

When we get home, Archer sits the moth on a countertop and starts researching. "Cecropia have wings up to seven inches," he announces while ransacking a junk drawer. He produces a measuring tape for positive confirmation. This moth measures more than seven inches across. It takes both of my hands—which, placed side by side, mirror the shape of moth wings—to hold it completely.

"If this is the largest moth in North America, what's the largest in the world?" Archer wonders. The answer: Australian atlas. He finds a cell phone app that makes it look like the exotic moth is in the kitchen next to us and holds it up next to the cecropia as if to measure the two against each other. The atlas, which shares similar warm-tone coloring, is no more magnificent than the cecropia moths that fly through these hills.

This moth is a reminder that we live in a temperate rainforest, a haven of biodiversity that, each year, slips a little further from being a sustainable harbor. On light pollution maps, the Appalachian chain is discernible, because it offers a slight East Coast swath of natural night that's hanging on, but just barely. I've heard that dragonflies with three-foot wingspans once zoomed through Ohio and West Virginia. Those creatures were alive when dinosaurs roamed the world. Yet night still holds and hides beauties like this. In the here and now, not some golden-day past.

As I try to figure out where we might put the moth, I remember a spare shadowbox that's been tucked into a closet. It is like the frames that natural history museums use to display pinned butterflies and moths. For years, I've used shadowboxes to frame found objects like river glass, delighted by the act of turning what might be seen as disposable into something of value by simply reframing it. Into this box, the moth's wings will fit like a puzzle piece. But how does one go about mounting an insect?

I look up as Archer jostles the piece of paper before suddenly jumping back. "Did you see that?" he asks. I nod. Our moth specimen just moved. "Alive, alive!" We take turns screaming enthusiastically.

Archer shifts the paper again. The moth responds by lifting his wings languidly. "What should we do?" Archer asks.

"Maybe we should offer some sugar water or a piece of fruit?" I venture.

We soon learn that cecropia moths cannot eat or drink. A cecropia moth's only option is to expend energy, pushing it toward future generations. "I wish we could take this moth to a wildlife center for help," Archer says.

But even in top form, giant silkworm moths like cecropias only live ten days or so. There's nothing for us to do. This is not a riddle to be solved. It's not a problem we can fix. When it comes to saving this moth, we are helpless, and it is an unwelcome sensation. It's also a tangible facing of the fact that, regardless of how we attempt to avoid it, none of us will go on living forever. We leave the moth on our kitchen counter, wings splayed across paper like watercolor.

~~~~~~~~~~

When I wake the next morning, there is a small part of me that believes we might find the moth flying around our house. But the cecropia—with his large, feathery antennae—is exactly where we left him. We leave him undisturbed until lunch preparations require relocation, determined to keep him as comfortable as we can.

The shadowbox comes with a backboard covered in black velour. I flip it inside out so that the box is not an enclosure but, rather, a pedestal. Then, gently as I can, I put an index finger on each side of the moth's orange-and-cream-striped body to lift him like a paramedic. The animal's legs are shrunken. But he is not yet dead. His half-moon markings look like celestial bodies, but scientists think they evolved to mimic the eyes of snakes or owls to make predators hesitate on approach. The patterns are protective.

Once the motionless cecropia is on his throne, I tap the backboard to see if he is still alive. He raises both wings, surprisingly strong and wide. Still unsure of where to move him, I put the inside-out shadowbox on my desk. This is the place I spend much of my waking life. It's away from the bustle of the kitchen and the low surfaces of my living room, where my dog might try to grab the moth as a snack.

Later, as I check email with the cecropia by my side, I sense the harshness of my office's overhead lights. I feel the blue spectrum of my computer screen. All the artificial lights of my office are hitting the moth's wings. There is maybe no scene I'm more familiar with than the one from my office chair. But suddenly, I'm seeing it from a moth's perspective, and it changes things.

I turn off my office lights. I turn off my computer screen. I leave my window blinds cracked, so that the cecropia will have the natural ebb and flow of day and night as miniature seasons experienced from an inside-out shadowbox that's balanced on my ink-jet printer.

I cannot sustain this creature. I cannot revive him with sugar water. I cannot do much of anything to ease his journey. As we rest there together, I realize that the only solace I can offer this cecropia is the darkness that he has been denied. But maybe, in an overlit world, this is no small thing.

~~~~~~~~~

I sit with the dying moth, day after day, night after night. I move from tapping the board every so often to waving my hand in front of his wings. I do not touch his corporeal being, but the moth senses me. I sweep my hands, churning air. The moth lifts his wings. It feels like sorcery.

Sometimes, when I'm close, my breath tousles his furry head, and it is enough to inspire a slight tapping of his half-moons against velour. I've had call-and-response communications with plenty of domestic animals, but not often wildlife. Then again, I'm beginning to think that's only because I haven't really been paying attention.

Human conversations with wild creatures are always happening. It's just that we don't realize it—so the conversations remain sadly one-sided. I've been unintentionally calling moths with artificial light all my life, but I've barely made out their responses, ignoring the fact that their actions had much to do with mine.

Aside from his shriveled legs, this moth looks fit. His wings, studied closely, remind me of wild rabbit fur, and his unruly orange head is akin to nothing I've ever seen outside of the Muppets. I grow to love him. Sometimes, when returning to my office after long hours away, I lean in and whisper, "Are you still here?" To this, the moth always answers with a gentle wing lift. We go on like this.

My call is always the same, but, in time, the moth's response begins to fade. Only once do I carry the moth from his perch by my office window, into full daylight, when family friends are visiting. The youngest among them holds out his hand, to measure the moth in relation to his own body, and asks of the giant cecropia, a creature wilder than he ever imagined might exist, "Is that real?" To almost everyone who encounters the cecropia, the concept that this massive beauty exists in close proximity unnoticed seems, at first, unbelievable.

Outdoors, this immobile moth would likely have already been eaten by a bird or blown into soil. I've removed this insect from the very cycles I long to connect with. But the moth has become an unexpected companion, deliverer of some wordless wisdom.

Each time the moth and I meet, he gently beats his wings against the backboard of the shadowbox, half-moons rising and falling. I try not to disturb him too often, and I take seriously my duty to guard his access to darkness. Incrementally, his wing movements weaken. It's uncomfortable to witness. It is, frankly, a little scary. I watch the moth's energy fade until, finally, his wings go dormant.

"Where did you go?" I whisper one afternoon, though I know there will be no answer. Only the outline of the cecropia's wings, the shape of two question marks, almost touching.

We have spent nearly ten days together, almost surely this moth's entire adult life. There happens to be a moon calendar clipped to my window blinds, so the cecropia's place of death is marked with hand-drawn moons etched as though on a tombstone. They map the nights that this moth lived. They show that, during our time together, the moon was continually progressing from a mere wink to a wide-open eye. While this moth with half-moons on his wings expended his last bit of energy, the actual moon—as seen through my office window—was reaching its height of light.

I cannot bear to pin this moth down like a museum specimen, so I place two pearl-tipped pins in velour to hold him upright, hooked under the crooks of his wings the way I looped my hands under Archer's arms to lift him when he was a baby.

I have seen live moths. I have seen dead moths. But never have I been through this process of watching life drain from a winged creature before. When I look at the moth's body, I cannot help but feel like his energy is hovering in the air around me.

I leave the cecropia alone for a few days, until one afternoon, when it feels right for reasons I cannot explain, I ceremoniously flip the backboard. The pedestal on which this moth has lived and died transforms, immediately, into a display case. There is still a slight impulse to take the moth out, to bury him, return him to cycles outside. But it is tempered by the memory of the child who asked, when this moth was alive, if he was real. I decide that the cecropia, in death, has things to teach by simply staying visible.

I stand there for a long time, holding the box of darkness, unsure of where to put it. I've always appreciated shadowboxes because they leave space for depth instead of pressing everything they hold against the hard limits of glass. How odd it is that we speak of shadows as voids when they are the very definition of a presence. Light flowing around something, the way ripples reveal rocks in a river.

Mothapalooza

My cecropia encounter would have been a gift at any time. But the arrival of this moth could not have come at a more auspicious moment. On the night Archer had his moth-centric revelation,

I was already planning a road trip to attend Mothapalooza, a moth-lauding festival held annually by Arc of Appalachia, a nonprofit organization with vast acreage under its protection in Ohio.

Every summer, the Mothapalooza festival erects larger-than-life lanterns. After sunset, these luminarias—some of which look, from a distance, like curved cathedral windows scattered through woodlands—lure moths from the sanctuary's dark forests so that moth-ers, the moth-loving equivalent of birders, might be able to get a look at them. I am now, more than ever, eager to join them, though my enthusiasm about drawing moths with artificial light has been tempered.

It seems like folly to travel up a mountain range to find what's surely all around me. But, as with so many nocturnal life-forms, I'm not only lacking in how to locate moths—I'm not sure how to relate to them. And Mothapalooza is full of people who've spent years considering moth-human relations.

As soon as I arrive, I run into an experienced moth-er, Jim McCormac, one of the festival's founders. He's loitering near a picnic shelter, where he's set up a light station that hasn't had time to draw much attention. It isn't so much the white sheet he's erected on the side of the shelter that draws my interest. It's that he's holding a gigantic UV flashlight.

He isn't haphazardly walking around with his light, as Archer and I did with ours; he's fixed his beam on the forest overstory, which is awash in purple light beams. When I walk over, he says, with barely an introduction, "I want to do a lot of looking around for cats tonight." He takes my silence as confusion. Which, admittedly, it is. "Caterpillars, I mean," he says. "I just call them cats." This is, I will learn, a classic moth-er reference.

Jim started Mothapalooza with a fellow enthusiast, and he can still remember his inspiration. He was sitting in a ditch when the concept came to him. He'd just found a moth. It might have been a cecropia, if he remembers correctly. Jim says, "I told the guy I was with, 'You know, if people could see how beautiful this is, everybody would love moths."

The first Mothapalooza was held in 2013, with nearly 200 people traveling from all over the country to attend the event. This year, with similar attendee limitations, people have arrived from Nebraska and Maine. Mothapalooza is so popular that, some years, it has a waiting list. Even after more than a decade of success, Jim cannot believe it. "When we started this, we had no idea anyone would come," he says. "We're really just a bunch of people who never grew up."

He taps the side of his giant UV flashlight and holds it out to me in official introduction. "This is the Beast," he says.

The Beast, as it turns out, is a brand-name UV flashlight that's a staple among moth-ers, and they tend to reference it by name. This Beast has its sights on an inchworm that's hanging from a leaf above us. It is too high for Jim to reach, so we stand there, eyes skyward, watching the inchworm tugging its slower half behind.

Jim comes across as equal parts biology professor and frat partier. It is, I suppose, the unique mix that led to Mothapalooza's natural-history party vibe. He's an expert in the biodiversity of nightlife. When I explain that I've been trying to get to know darkness via the creatures that depend on it, he nods in solidarity. "I've come to appreciate night as a whole other world," he says. "I've literally spent thousands of hours outdoors after dark. You get used to darkness. When you can name the sounds you hear, they're not so scary. Moth-ing is best in the witching hours, midnight to four in the morning. Who can stay up? Even I can only do it sometimes. But doing it sometimes is all you really need to get an understanding of what's going on out here."

It was plants that first got Jim excited about nature. He has been a botanist for most of his career. At one point, he collected a glade mallow plant, which was rare in Ohio, his home state. He took a sample of the species to press dry so that it might be displayed in a natural history museum. In his office, during preparations, he found a caterpillar on it. "Up to that time, I'd seen a lot of caterpillars," he says. "Usually, I'd just flick them into a trash can; I didn't want them to mess up my nice, pressed plant, you see. But this time, something made me pause. I took that one to an entomologist friend, and he said, 'I can't identify this as a cat.'"

Many moths are hard to categorize in caterpillar form; it's hard to tell what kind of moth they'll become. Jim's friend agreed to keep the caterpillar to see what might be revealed in time. "A while later," Jim recalls, "I got a call from him, and he was like, 'You'll never believe it, but this morphed into a bagisara!' It was only like the third one ever found in the state of Ohio. And it all started making sense: rare plant, rare caterpillar." He realized then that he hadn't been seeing caterpillars for the leaves. And neither could exist without the other. It was, for him, an awakening.

Kim Bailey, a first-time festival attendee, has sidled up to us in the dark, drawn in by the Beast and held by her curiosity about botany. She's a milkweed farmer who, like me, has traveled here from the mountains of North Carolina, and she's always had an interest in the way plants and butterflies collaborate.

Kim started her milkweed farm to support monarchs, who famously travel vast distances for their migration and depend on milkweed for their reproductive needs. In recent years, she has planted pawpaw plants for zebra swallowtail butterflies, black-and-white-striped creatures that depend on native plants that have lived in these mountains since mastodons roamed them.

If you see a pawpaw tree, it's a clue that there are swallowtails nearby. Just as, if you plant milkweed in the monarch's Appalachian flyway, you will likely have those winged visitors when they're en route to and from Mexico. But even though she's spent her adult life fostering plant-butterfly relationships, Kim's introduction to moth-ing came unexpectedly.

Nocturnal pollinators weren't even a distant thought when she planted night-blooming primrose flowers in her garden. Kim says, "I'd heard primrose bloomed at dusk, and I wanted to witness that." So, she did, and because she was so focused on seeing that flower bloom in real time, she started visiting it, her schedule aligning with that of moths who'd come to visit. One of them was a hawk moth, otherwise known as a sphinx. It was as large as a hummingbird.

"I never knew something like that existed," she said. "Every night, something new and more exciting came to those flowers." For a woman who'd dedicated her entire life to flowers and butterflies, it was like a sudden doubling of what she loved most about the natural world.

"Day shift, night shift," Jim says.

A few years ago, Jim partnered with an educator to write a book about moth gardens. It's specific to native species in his part of Ohio, but it's one of the first titles to ever focus on the connection between nocturnal bloomers and moths. Given the proliferation of titles about butterfly gardens, it seems to address a tremendous oversight. Because there are more than 12,000 moth species in North America—and roughly 800 butterfly species.

For all the talk I've heard about butterfly bushes, I'm not sure I've ever heard anyone chatting about how to attract moths. A quick internet search will confirm that, mostly, when we talk about moths, we're talking about the most effective ways to kill them. Even Kim, who has spent a lifetime cultivating plants for butterflies, has only just begun to pay attention to the winged workers who tend her land while she's dreaming. But even when we don't recognize moths as important, we still depend on them.

According to the USDA, approximately 80 percent of the world's flowering plants require pollinators to reproduce, including three-quarters of agricultural crops. A 2021 study on nocturnal pollinators that was conducted by the University of Arkansas led researcher Stephen Robertson to call moths "unsung heroes of pollination," indicating that, as an under-studied group, moths could potentially beat out butterflies and bees in pollinator importance globally. But just as we're discovering how important moths are, they're disappearing.

In 2021, the *Guardian* reported that the insect population, including nighttime pollinators, has declined in abundance by 75 percent in 50 years. For many people—including, thus far, me—those losses have been abstractions, because we haven't known what we had to begin with. And even though moths still swirl after dusk in every state of the union and throughout diverse global regions, many people don't know about their local moth populations or how light endangers them.

Studies have shown that artificial light impairs moths' ability to pollinate by as much as 62 percent. Verdant landscapes depend on them, but moths' alliance with darkness has given them a dull reputation. Understandings of moths seem to be limited to the idea of them as shadows beating against lights or marauders eating coats in closets.

But moths are beautiful far beyond their reputations. They have patterns of stained glass and textures resembling everything from cotton balls to feathers, often with butterfly-bright colors. There are more than 160,000 recorded moth species in the world, and new-to-science moths are found all the time, many representing plant species they coevolved with, their wing patterns serving as clues to what else lies in darkness. Some, like the cecropia, indicate the presence of cherry trees; others, jewelweed. They come with names like pistachio emerald, scribbler, and green marvel. They can be as small as dust specks and as large as baby songbirds.

Tonight, Jim, supreme moth partier, wants to help us absorb the magnitude of it all. Just uphill from where we're standing is a station that he claims to be the most consistently awesome show at Mothapalooza. "We should go up there and see what cats we find along the way," he suggests.

As we ascend, we're joined by other people who've been drawn to our group by the Beast. Every budding moth-er seems curious about UV light being used as a cat-hunting device. Our group of three, then four, then five moth-ers soon grows to over half a dozen people. Jim is like a pied piper. Only instead of using a flute to put us in a trance, he's drawing us in with UV beams.

The incline of the hill isn't steep, but it might explain why the light station we're headed to is of particular interest. Moths are known to demonstrate hill-topping behavior, a push for elevation. Apparently, hilltops are natural nightclubs of the insect world, where moths gravitate upward to increase their likelihood of encountering mates.

They use topographical clues to move upward until they reach summits. The ones who make it to the peak are thought to be considered particularly fit and attractive, rugged as rock climbers. Even slight inclines can encourage this behavior. To moths, even mole hills are mountains.

When Jim's phone lights in his pocket to announce a call, he immediately shuts it off. "We don't want to confuse the moths," he says. "When moth-ing, you want to create a fixed point of light as a focus." We all agree that too many lights would also distract us.

The overstory is dotted with caterpillars that shine white through the Beast's purple haze. "Here's a sycamore tufted," Jim says, tugging the limb of a tree so that we can inspect the caterpillar closely. "They pretty much only eat sycamore leaves."

Kim has a magnifying monocle around her neck. She lifts it to her eye to inspect the caterpillar's spikes, which appear as spun glass. In the moment, both Kim and the caterpillar look like characters straight out of Alice in Wonderland. These caterpillars can be found falling out of sycamore trees all over New York City's Central Park, where bands of moth-ers probe the city's reservoir of darkness.

It would be easy to view those urban moth-ers as extreme risk takers. But, while it can unquestionably be a dangerous place, Central Park, statistically, reports some of the lowest crime rates of any police precinct in the city. Having more people around, like having more lights around, doesn't always make things safer. When I share this bit of trivia with Jim, he shrugs and says, "The whole world is full of surprising realities once you're alert to them."

Many species of caterpillar—be they destined to become butterflies or moths—are nocturnal in their early life stages, hoping to protect themselves from birds. On a single summer day in Ohio, red-eyed vireos alone eat roughly 30 million of them. This is impressive when you consider that some caterpillars, like hickory horned devils—which, if they survive, become regal moths—can be the size of cocktail franks.

Jim seems to have vireo-level skills, noticing leaves with telltale holes in them from ten feet out. Often, these holes appear as damage, but when caterpillars have evolved with native plants, it's a relationship that does no irreparable harm. There are exceptions—destructive exceptions, mostly tracing back to some non-native plant introduction and human interference—but, in general, holes in the leaf of a healthy native plant are a hieroglyphic script of how species have learned to coexist.

A garden with untouched leaves might be prized, but they are—from a caterpillar's perspective—mausoleums of isolation, whereas native plants with chomps taken out of them are celebrations of interconnectivity. An imported plant that has no relationship to native insects might be completely intact, but its lack of holes means that it lacks a sense of belonging. Plants were meant to be touched, and nature is a collective art project that we're all participating in—whether we like group assignments or not.

Without caterpillars turning plants into digestible protein, bird populations would perish. And without the specific plants they depend on, caterpillars would disappear. And without the moths the caterpillars become, plant pollination would be stunted. The connections are seemingly infinite—and complicated. When I've come across a caterpillar throughout my life, I've automatically thought I was seeing a future butterfly. But, given the vast outnumbering of moth species, as it turns out, I've almost always been looking at a moth-in-progress.

Jim seems pleased with the variety of life that's clinging to the underside of the sanctuary's leaves. "The sheer abundance of cats is pretty much what makes the world go round," he says.

~~~~~~~~

At the top of the hill, an arched doorway of light appears in a forest clearing. When we get closer, we see that it's a piece of metal, bent and stretched with a sheet to form a canvas. The contraption is, on both sides, illuminated by a mercury vapor lightbulb sitting on a tripod.

Moths' attraction to the moon is thought to be due to something called optical infinity: a constant light that's far enough away, at a constant angle, to help sustain flight in a one-way direction. Artificial light is, to moths, confusing in part because it calls from a variety of angles. This sheet-stretched contraption solves the moths' endless merry-go-round issues by providing a place for them to land. The creatures on this moth-ing sheet shot for the moon and ended up finding a human-built one, right here, on this Appalachian hill.

"Io," someone shouts as a moth lands on the faux moon. The moth's odd name—ode to a princess of Greek mythology—comes out like a strange battle yelp, "Eye-oh!"

The female io is primary yellow with what appear to be two

blue eyes on her wings. "Well, hello there," someone says to the moth in greeting. Then, to no one in particular, they ask, "Isn't she beautiful?"

She is beautiful. And, though she has snake-eye patterns, she appears to be staring without menace. "Hey, another io! A fresh one! Over here!" shouts a voice from the other side of the sheet.

"Fresh? What does that mean?" asks a woman who has moths dotting her pants. Moths are landing on all of us, but her outfit—white as the moth-ing sheet—looks like a covered canvas. She's already explained that, when she chose to wear white, she didn't know it would have moon-goddess consequences.

"When a moth is fresh, that means it's just emerged," Jim explains.

By the sheet, a few feet from where we're standing, someone notices a newborn regal moth and starts squealing. Regal moths look like royalty. They're sometimes called walnut moths, so deep is their abiding connection to walnut trees. Regal and cecropia moths belong to the same family. It's strangely comforting to see this regal moth's fluff, orange and unruly.

Before I notice what's happening, Jim has lifted the moth from the sheet. He holds the insect up to my arm for transfer. Her wings are immobile, but her legs seem to be strong. When she moves across the thin skin of my inner arm, I shriek. "You're okay," Jim says, voice steady. "Some cats can sting, but moths aren't going to hurt you. You're safe. You learn when you touch things." Or, in this case, when they're touching me.

I am not afraid, per se; I'm uncomfortable. Moth legs against my skin is a strange sensation. I feel exposed, vulnerable. Even though, consciously, I know it is the moth that's endangered here. It's strange how often humans feel like prey when we are, in fact, the predators wreaking havoc—often unwittingly.

This moth is exploring my inner arm as if it is an offshoot of the moon, but that doesn't mean she's here to be petted. I'm careful not to touch her wings. If I did, my fingertips would be coated in shimmering scales like eye shadow. This would not kill the creature, since moths are always naturally shedding scales, but it might cause premature aging.

The woman crouched beside me clearly knows her moths, and my shriek has caught her attention. As a preschool teacher, she is around children who are having new experiences pretty much every day. She says, "People are afraid of moths because they're afraid of nature. A lot of my job is teaching people how to relate to things without bias."

Not long ago, she helped an adult friend overcome a serious hesitation about moths. Even people who love butterflies can find moths intimidating. It took a few months, she says, but she introduced him to one moth species at a time, building up his immunity to fear. "When he saw how excited I was about moths, and how beautiful they can be, he started to change his mind about them."

This is, at root, the premise of Mothapalooza. Apparently, it's working. This hilltop station has drawn half a dozen giant silkworm moths at once. And more and more humans are scaling the hill to watch them frolic.

Gently, I put my arm against Kim's arm, so that she might carry the regal moth for a minute, and I wander over to inspect a snowball-looking moth that's landed. "A white flannel moth," someone announces. The moth looks like a miniature stuffed animal with yellow antennae. "Look, from the front angle, this looks like a panda bear!" Jim says.

"Gah, I love this one!" the preschool teacher exclaims. "And white flannels aren't even in my top ten favorites!" She stands back, waiting for me to take the lead in a human-moth interaction. But I keep my hands to myself. So she applies a little pressure: "Pretend you're going to touch their nose and they'll climb right on." When I hesitate, she demonstrates by putting her index finger to the moth's face.

I haven't formulated a list of top ten favorites, but, as much as I love the panda now balanced on her index finger, I'm already distracted by a rosy maple moth that has landed on her fanny pack, with wings the pink and yellow of summer lemonade. This moth's placement is notable since the creature has come to rest directly next to an enamel pin—a perfect artistic rendering of a rosy maple moth.

When I point out the pin's living twin, the teacher laughs. "I'm wearing matchy-matchy moths!" The moth has likely mistaken the enamel version of itself for the fruit of a red maple tree, the rosy maple's favorite species. It is those exotic-colored tree bits that this moth has evolved to mimic. In daylight, the moth needs to hide from songbirds, and it can, thanks to camouflage. Rosy maples look out of place everywhere, except for with the tree to which they belong. Among that tree's fruit, the moths become invisible, even in full daylight. Next to this pin twin, the moth is trying to hide.

One of the only downsides to hanging out at the hilltop light station is the actual light, which looks like the flash component of an old-timey camera. When the light station attendant notices me wincing under its power, I mention that it seems worrisome to be calling moths with light—even though, unintentionally, we all do it, all the time.

He assures me that it's something that he, too, considers every time he goes moth-ing. "If you kept doing this all the time, you might create issues for moths. But this is limited to two or three nights at most," he says. "Even outside of this event, moth-ers are usually mindful of what they're doing and how often they're doing it. Using light like this is a way of training yourself to be aware of it."

Mercury vapor are some of the most popular bulbs used for moth-ing, but different wavelengths of light seem to attract different species, something that's not well understood in terms of physiology. And the apparent attraction varies with individual moths. The attendant points to a regal moth. "There are probably ten more of these nearby. They're around, but they're choosing not to fly over."

It's surprising to learn that moths, generally characterized as light-attracted zombies, might act on free will. Maybe it is not unlike how, to some humans, this scene is like something out of a horror movie they'd never want to see. And, to others, it is the epitome of beauty, something worth driving across five states to witness. This, too, is difficult to explain.

More than one of the seasoned moth-ers on site has seen people go from viewing moths negatively to positively in the span of a single evening. Rosy maples tend to be what Jim calls a "spark moth," a creature fantastical enough to make someone who has randomly spotted it ask: If that sort of Dr. Seuss character exists in my yard, what else might be out there? At Mothapalooza, it's hard not to ask this every other minute.

The later it gets, the more moths join us, until the hilltop is a

cyclone of wings, every flutter a brush stroke against the moth-
ing sheet. There are dozens of species intermingling, all with their
own patterns, their own colors. Despite the living confetti falling
against my arms and my legs, my attention is drawn upward,
toward the tops of the tree line, where something is swooping.
The animal is the size of a bat and the color of a star, wings twinkling in mercury vapor. Slowly, an imperial moth—another of
the cecropia's silken family—comes into focus, spinning through
this midsummer sky before ultimately landing. Unbelievably,
this moth is followed by another, similarly large imperial who
comes in hyper, wings beating and beating.

"He'll settle in a minute," Jim assures us. "This light's just
too exciting."

Other giant silkworm moths that have chosen to land outside the sheet's perimeter have begun to tremble, their wings
shuddering as if in fear. "The big ones have to warm up before
they fly," Jim says, reading their behavior as a signal that they
might be attempting to leave. It inspires him to do the same, and
given that he's the one who led us here, we have an instinct to
keep following him. Apparently, there are other moth stations of
enticement. He clearly knows how to navigate this party. Still,
we don't leave immediately. Every minute, something unexpected
flies out of nowhere, and, increasingly, it's not intimidating; it's
exhilarating.

"I'm having a hard time leaving one light to go to the next
light," Kim says. I feel the same. How in the world, we wonder,
could anything top what we've just seen?

When we finally catch up to Jim, he's standing in front of
a station that lacks the dreamscape quality of the hilltop light
portal. It is just a bedsheet hung over the back of a Subaru

hatchback. But the giant leopard moth that's landed on it makes everyone ignore the station's lack of refinement.

I've already seen almost one hundred species, but the giant leopard might be the most stunning. The moth's stark white body is covered in thick black circles. Some appear hollow; others full of a blue-green iridescence.

"What color would you call that?" someone asks, marveling over its richness.

"Iridescent blue-ish?" a group member says, dreamily.

We're in rural Ohio, communing with leopards. It's preposterous. I'm swimming in awe even before the moth reveals an abdomen of black and brass, flashes of lapis blue and turquoise green that remind me of Egyptian jewelry, sacred scarabs.

We develop such fervor that Jim suggests we should give the moth a little space. It's only then that many of us realize we've been fawning like fame-crazed teenagers. One by one, we move away, careful not to crush the moths who've landed on the station's white sheet, which is running along the ground like the train of a bridal gown.

Throughout the evening, many leopards will reveal themselves here. There will be moths more diverse than we can imagine. But we, new to moth-ing, still don't fully think of night as a place of abundance rather than scarcity. We're acting like this is our only opportunity. We see this leopard as something peculiar, but part of its beauty lies in the fact that it is so very ordinary.

Giant leopards are partial to dandelions. Those are the plants that call to them, strong as a light source. And, despite the leopards' seemingly uncommon beauty, like the weeds they depend on, they are not rare. Giant leopards can be found in Canada and Florida, from Minnesota to Texas. Every dandelion left to grow in a yard has the potential to bring these creatures closer; every dandelion killed in favor of grass pushes them away.

Jim has seen countless leopards. Unlike us, he understands that night is full of astounding creatures that will be pouring into this light island for hours. He shakes his head, amazed not by the appearance of this moth, but by our exaggerated reactions to it. "This would be like finding a cardinal for birders," he says, citing one of the most common birds of the East Coast. "Giant leopards are everywhere."

~~~~~~~~~

Around midnight, I visit a sheet that's been hung on the side of a utility barn, where I run into a brother and sister who've traveled across the country to attend Mothapalooza. "I had a favorite moth species, then it changed. There are just too many amazing ones," she says, eyeing a newly arrived red-and-gray moth.

"Scarlet lichen moth!" I blurt.

Jim overhears. "Man," he says, "you've learned a lot tonight."

It's hard for me to remember a period of my adult life when I've absorbed more in such a short time. A few hours ago, I was unfamiliar with the lines of a scarlet lichen moth's wings, but since being introduced to the species, I've helped more than four other attendees identify it. Now that I recognize its pattern, I cannot unsee it. And my field of vision is expanding with every winged arrival.

As the crowd thickens, so does the moth cloud that's been collecting. There are so many moths, in fact, that some people are wearing masks so they will not ingest these airborne life-bearers. I keep my mouth closed, out of necessity. I take note as mothers make identifications. There's a bold-feathered grass moth. A basswood leaf roller moth.

When the station attendant sees that I've started spending time with the tiny moths rather than hovering around the larger species, she walks over and says: "Silk moths are the bling of moths, but these micro moths, these are the ones I like best. Most people don't pay attention to these little specks. But, if you really look, they're like glitter."

"Glitter sounds like bling," I say.

She laughs. "I guess you're right. Micro moths are just my bling of choice!"

Her commentary inspires me to lean in, side-mouth breathing so that I don't inhale wings. From a distance, these moths might have appeared boring. But, dialed in, I can see that many of the small moths are golden. And I don't mean yellow. I don't mean the glint of plastic sequins. I mean the gilded gold of cathedrals and holy statues.

In Japan, Buddha statues have historically been gilded not to flaunt wealth but, rather, to catch the interplay of candlelight's natural ebb and flow in windowless spaces. Sparkle is, after all, about how light submits to darkness—and darkness to light, in turn. These moths are crowns of gold, halos hovering above and encircling our heads like the famously metallic halos of Greek Orthodox paintings.

I'm so immersed I hardly notice that Mothapalooza's roving golf cart has approached the station until Jim suggests that Kim and I, and a couple of the other moth-ers we've been walking around with, should climb aboard. He's been keeping tabs on what's going on elsewhere via walkie-talkies since the sanctuary is too remote for cell service. "The moth station I'm taking you to now is the hot new spot," he says. "I heard they've got a couple of sphinx moths."

Some people go club-hopping for fun in a city; some people

get their kicks from witnessing the biodiversity of moths in Appalachian Ohio. If Mothapalooza is any indication, the latter attracts a surprisingly large crowd. I squeeze into the cart with Kim and her friend Sharon, all of us facing backward. When we take off, the light station fades behind us. The road is completely tree-lined. It's like we're traveling through a tunnel with light at the end. Only we're being pulled into darkness.

It's a strange sensation, watching light turn into a memory. When I point out the oddness to Sharon, she laughs. "You're right. We're not headed toward the light at the end of the tunnel. We're going the wrong way!" But to reach whatever's next, we'll have to loosen our grip on what came before. It isn't easy. This we've learned at Mothapalooza, over and over.

I wonder if it is akin to what a caterpillar feels when going through the alchemy of transformation, retreating from light into darkness for rearrangement. It's a fleeting thought that soon gains resonance because, though there is no way for me to know in the moment, we're on a journey toward actual metamorphosis.

Minutes later—away from the glare of the next party-hopping moth station—an unidentified caterpillar pupates on the side of an Arc of Appalachia guesthouse. One minute, the caterpillar is there. The next, he's gone. Only he hasn't vanished; the creature has just turned into something completely different. The moment marks both an end and a beginning.

Jim, previously unflappable in his boisterousness, turns solemn over watching the unidentified caterpillar shake as though preparing to explode before seemingly turning inside out. With surprising speed, the caterpillar transforms from a creeping crawler into a seed of dark mystery that Jim—even armed with the UV Beast—cannot inspect. "I'm a scientist," he says. "But when you see something like that, it's hard to call it anything other than magic."

~~~~~~~~

There are plenty of activities to enjoy in daylight hours at Mothapalooza as everyone waits for the prime attractions that can be found every evening, but I, for one, find myself too exhausted to join a guided hike or creek exploration after staying up late. So the next day, I spend time in the cool interior of the nature center, where I find a display that's been set up by The Caterpillar Lab, an educational nonprofit that features live exhibits notable for using microscopes to project the world of caterpillars onto movie screens.

A crowd has gathered around the daily presentation. Seeking a little space, I wander over to a buffet in the cafeteria that's lined with vases that hold not flowers but an assortment of tree branches. And on those branches there are armies of caterpillars.

I arrived in Ohio half-hoping to spot a cecropia in flight. But I've learned that it's a little too late. Like fireflies, the cecropia species' flight season is fleeting. Yet, here, I improbably have come face-to-face with a group of cecropia caterpillars.

The green cherubs are dotted with primary red and blue, and their feet are wrapped around twigs in gummy-bear hugs. I am so close to them, I can see their mouths churning leaves, tree energy that they will turn to silk, fantastical as spinning gold from straw. After nursing a moth that had no mouth—just like many, though not all moths—it's inexplicably meaningful to watch these cecropia caterpillars eating. This is what goes into every moth's making. This is a minuscule action that shapes vast landscapes.

My caterpillar-communing is interrupted by a boy, maybe four years old, who has come to stare alongside me. A Caterpillar Lab assistant comes up and tells the boy that the cecropia caterpillars, unlike some of the others, are okay to touch. Still, he hesitates, and his mother explains that he was recently stung by a hickory horned devil, so he is nervous.

Some cats are venomous, and some who are not venomous can still trigger our immune systems. Some people don't react at all; some are very allergic. Like with any allergen, things are unpredictable and vary person to person. Reactions are dependent on personal chemistry.

"Do you want me to touch the caterpillar and let you know what that feels like?" I ask. He confirms with a nod.

The caterpillar's body is surprisingly buoyant. Young and vibrant, this creature's moth days are still ahead. The caterpillar's setae, pale filaments, are prickly, but not unpleasant. I turn to the boy who is awaiting my report. "That felt like the bristles of a hairbrush," I say.

The boy buries his face in his mother's skirt, peeking every so often at the caterpillars. His mother inspects the branches and is pleased when it's confirmed that they're black cherry. "Oh, we have black cherry in our neighborhood," she says, clapping. "We might have cecropia!"

There is, at the far end of the buffet, a collection of cecropia cocoons. With the blessing of Caterpillar Lab staff, I lift one and turn it in my hands. It feels both tender and tough, coarse silk protecting an inner layer of fine insulation.

Caterpillar silk was a key component of the earliest bulletproof vest, which saved its first human life in 1897. Today, researchers are working to use the material of silkworm caterpillars for medical bandages and ever-stronger bulletproof materials to protect humans from artillery of our own making.

This cocoon looks like the aftermath of a leaf's death, but it is shrouding the cecropia's most vibrant life stage, one that will arrive only after it endures the coming, colder season. The silken bag indents when I gently pinch to feel the pupa it's carrying inside—a hard cylinder, shaped like a bullet. If I'd seen this on the ground, even ten minutes ago, I wouldn't have recognized it as anything special.

~~~~~~~~~~

The Caterpillar Lab's founder, Sam Jaffe, has shaggy hair and a way of making even the most unattractive of caterpillars seem charismatic. The crowd around his microscope never dwindles enough for me to meet him, but I run into him later in the day when he's on a break, browsing the edges of a parking lot with a red shopping basket. Small plastic vials roll around inside of it, clanking every time he leans down to inspect something.

When he sees me, he offers an unprovoked explanation: "The people who come here, I think of them as story seekers. So I'm out here finding some new stories to share."

He's on the hunt for leaf miners that he might later put under the Caterpillar Lab microscope. They're tiny caterpillars who, in larval stages, eat the tissue of plants from the inside. "Moths have a lot going on. People coming to the station sheets at night forget that the stories they're seeing are much bigger than that moment. Leaf miners spend weeks eating individual cells. Under a microscope, we can watch this happening. The story of these creatures extends beyond nighttime," he says. "Every little spot on a moth-ing sheet has a magnificent life history. When you see a little dot, you're witnessing an opportunity to ask questions: Where did this come from? What did it need? People obsess over moth species names, but we can't really know what anything is if we don't understand where it began."

Gardeners are, as much as anyone, familiar with caterpillars, which I've heard some of them refer to as the bane of their existence. When Sam travels, he sells moth-loving bumper stickers. When I quote some of the ones I noticed in the nature center, he chuckles. "*I Grow Tomatoes for the Hornworms*—did you come up with that tagline?" I ask.

"Yeah," he says. "I know people who decide to plant extra tomatoes to take the caterpillars into account instead of getting mad and trying to fight them." Humans often focus on what moths are consuming. But moths and plants do not have one-way conversations. Garden-moth relationships are complicated. Where there are moths to take, there are often moths to give.

At the mention of another bumper sticker—*Moths Are Better*—Sam falls into a full laugh. "All those bumper stickers are a little sassy," he admits. "But butterflies get all the credit. Moths need people standing up for them. Butterflies are just a type of moth, really, scientifically speaking."

The woolly bear is arguably the most famous caterpillar in the United States. When I bring up his *I Brake for Woolly Bears* sticker, Sam asks if I have heard about their weather-related folklore and the festivals thrown in honor of them, and it's my turn to laugh. There are a few festivals in the United States dedicated to the species, but the largest takes place in the North Carolina mountains, one county away from my house. I've traveled to Ohio for a moth festival, thinking the concept was outlandish, when my own community throws an annual party for caterpillars. "Where I'm from, we call them woolly worms," I say.

Woolly worms are found in Mexico, all over the United States, and into Canada. They're one of the most identifiable caterpillars in North America. The weather predictions he's referencing are based on the idea that the size of a caterpillar's stripes indicate the harshness of an incoming winter. It's Appalachian folklore I've known all my life: If you see woolly worms that are mostly milky brown, there's a mild winter to come. If the caterpillar is mostly black, the winter will be hard.

More than forty years ago, a local entrepreneur—thinking of Punxsutawney Phil, the groundhog who famously predicts whether there will be more winter or an early spring by the status of his shadow—determined there was something of interest in these caterpillar stories, passed down for generations in the mountains. If economy-boosting tourists could get excited about a groundhog's shadow, why not promote the fortune-telling stripes of a caterpillar's back?

The resulting, circus-like Woolly Worm Festival was a resounding success. Each year, 20,000 people visit my home region to attend it, and roughly a thousand of them bring caterpillars to race up strings. There is always some variance in caterpillar stripes. So the fastest caterpillar is declared the official oracle of weather.

I have, like most people, thought of woolly worm predictions as quirky Americana fun. Many entomologists also laugh it off. But studies suggest that the length of a woolly worm's brown-and-black segments are indicators of whether they got an early or late start the previous spring due to temperature change. So though they have not been proven to know what the future holds, their colors do scientifically correlate with the previous winter's weather. History is by its very nature recorded backward. And woolly worms might not be prophets, but they are, apparently, climate archivists.

When I tell Sam about the excitement of people racing their woolly worms—cheering for an early summer or an extended ski season—he nods in understanding. He's seen adults and children alike shout encouragement to a caterpillar pupating under his microscope, invested for reasons beyond their ability to articulate. Sometimes, people cry. The emotionality of it has inspired him to look for mythological stories related to moths. The dearth of findings has been disappointing.

"It's surprising to me that more cultures don't have moth mythology. Western cultures are especially lacking," he says. "Imagine if you were a prehistoric human and you had a cecropia flying at you. I feel like that would be something you'd want to talk about. But maybe it's just that nobody pays attention to moth stories." Or, maybe, prior to electrification, moths and humans didn't have as many close encounters.

Any conversation about moth mythology would be remiss without a mention of Mothman, so I ask Sam what he thinks of Appalachia's famous cryptid creature, said to be half-moth, half-man. I have a feeling it isn't the type of ancient story Sam has been seeking, but I'm not convinced it's unrelated. As a New Englander, he's unfamiliar with Mothman, but he finds the idea interesting.

"A moth that's part human?" he says, trying to reconcile the concept.

I drove right by a highway sign for the Mothman Museum in West Virginia on my way to Mothapalooza just days ago, shaking my head at the chances of traveling up a mountain range to attend a moth festival and having my route take me right past Mothman's hometown of Point Pleasant. I've always thought of Mothman as a quirky relative of Bigfoot. But the wonder dawning on Sam's face makes Mothman's story—which is familiar to me, strange to him—seem like Appalachian mythology with Greek gravitas.

As the story goes, in the 1960s, a large-winged creature was seen by some Point Pleasant grave diggers and, later, by a group of teenagers. Mothman was said to be a larger-than-human, black winged creature with red, glowing eyes, a monster that chased the teenagers from an abandoned World War II munitions area—an 8,000-acre property that, to this day, reputedly holds bunkers of unstable explosives.

People have hypothesized that Mothman has been, from the beginning, a misidentified sandhill crane or a large-bodied owl. Others say maybe it's one of those animals, mutated from the toxicity of a nearby factory. Still more wonder if aliens landed, or if the government was testing covert flight equipment. Despite all the descriptions and theories, Mothman has never been reported as directly harming any witnesses. In fact, historic sightings often feature people questioning if what they'd encountered was a benevolent prophet or a bringer of destruction.

In the 1970s, *The Mothman Prophecies*, a book about the cryptid's reported appearance just before a local bridge collapse that killed 46 residents, gained a global reputation. The book went on to be adapted into an early 2000s movie that sealed Mothman's status as world-famous, with almost all related story-lines examining whether he is a hero or villain. Today, up to 75,000 people from around the world visit Point Pleasant annually looking for Mothman. They come to visit the museum that's dedicated to him, lining up by the dozens, even on Sunday mornings, filling the otherwise empty streets of Point Pleasant as they wait for the museum's doors to open.

Mothman and Mothapalooza fans rarely cross paths—which is, maybe, why Sam has never heard of Mothman, despite being in the caterpillar business—but if they did, they'd find that they share a penchant for moth-loving memorabilia. The Mothman Museum sells magnets that say things like *Fear the Dark in Point Pleasant* under images of a hulking monster. Mothapalooza, too, offers souvenirs. One of its popular T-shirts reads *Welcome to the Dark Side* under the outline of a hand-drawn moth, delicate and lovely.

It's a tale of two moth search-party invitations—one to explore the nocturnal biodiversity of moths, some still undescribed to science, the other an invitation to explore night's doom-and-gloom reputation, as depicted in cryptozoology. Juxtaposed, they seem like an ink-blot test of how people view night. And their parallel, dark-centric slogans inspire me to ask Sam, "Do you think people's fear of moths is related to a fear of darkness?"

In some ways, it seems that people think of butterflies in contrast to moths, like a sort of Dr. Jekyll and Mr. Hyde situation, with night acting as a nefarious catalyst. Sam sets his grocery basket down. "You know," he says, "I haven't really thought about it before, but I don't think people's negative feelings about moths have as much to do with darkness as they do with light."

I can imagine no T-shirt—in naturalist or cryptid circles—that reads *Fear the Light*. "What do you mean?" I ask.

"Well, most of the time, when people encounter a moth, they're meeting a moth that's disoriented by light. We don't often meet moths on their own terms; we engage with them in chaotic situations we've created."

Even now, after experiencing the overwhelming diversity of moth life in Southern Appalachia with moth enthusiasts, I'm not sure that I have ever encountered a moth in natural darkness. When I've seen moths, they've always been lured by something—a UV light, a flashlight, a porch sconce. The reality is that pretty much every moth I've ever seen, even here, at a festival celebrating them, has been in crisis, stressed by artificial light.

"People can't handle the unpredictability of moths in disoriented states," Sam says. It is patently unfair, thinking about this now, that we provoke moths and then blame them for acting erratically. Then again, it doesn't seem all that different from what we, in the technological age, do to ourselves, as humans.

"Those moths, they just want to be doing their own thing in the dark," Sam says. "In that state of artificial-light disorientation, it feels like a moth's attacking us. But moths approaching us in artificial light just don't know where to go, because we've put giant lights in their faces."

Around us, leaves start softly clapping in the breeze. With new perspective, I imagine all of them full of tiny leaf miners, swinging from limbs, just going about their business. Lepidoptera are considered amazing because of the metamorphosis they go through to gain wings, but they are always in the process of shape-shifting landscapes. Forests and fields, yards and gardens—from the viewpoint of a caterpillar, they're all like scenes from the interior of a gigantic lava lamp, shapes and colors shrinking and swelling as they chew their way to change.

We wade farther into weeds. With each step, grasshoppers

pop up in greeting. Sam singles out a leaf that's encasing a hungry caterpillar, and he puts it in his basket. In a few hours, he will showcase this leaf miner eating cells on stage, its micro-level work broadcast onto a movie screen in front of hundreds of people. This caterpillar, plucked from obscurity, will soon be a bona fide celebrity here, in Mothman country.

Leaf miners, as it turns out, are the caterpillars that go on to become micro moths, glitter bling. Some of the species I saw last night likely had their start not on a leaf, but in the darkened interior of one, writing scripture that can be seen—with my naked eye—on a plant's exterior. "My job," Sam says, "is to remind people that there's always more to the story."

Invitation to a Moth Ball

A *few days after I* get home, I hear from Sharon, who has already collected black lights and ordered a mercury vapor bulb for her own backyard, located a few hours south of mine. She's started inviting friends over to test the equipment at events she's referring to as Moth Balls—fancy dress optional.

People might put mothballs in closets to deter the few species that feast on clothes. But moth-ing is full of bling more deserving of formal galas. "I'm addicted now," she says. "I can't wait to get back out there every night. There are surprises every time!"

I've already been scouting my own moth-ing equipment. I don't have a mercury vapor bulb or a tripod, but I do have UV lights and a few high-power flashlights. A few days after I get home, I wrap a white sheet around my body like a toga and walk down to the river alone. The cotton grows heavy as it brushes wet grass, cool against my ankles.

The temperatures do not bode well for a good turnout. But I'm too curious to delay. There is, because of my time with the cecropia, a lingering sense of guilt in purposefully setting out lights, though I'm doing it to gain awareness of what my dark yard holds. In general, if darkness is not re-centered as a force that nurtures and harbors life, if we continue to talk about it as a deadened state, it's doubtful that we will ever be able to shift the cultural conditioning that has led to a broad-spectrum lauding of light and the vilification of darkness. No one nurtures what they view as worthless.

I hang a piece of twine between two locust trees, hopeful that forests across the river are prime moth habitat. I am only now beginning to understand that light pollution is particularly noticeable to me because, in these mountains, darkness has not completely fallen to artificial light. I live in a frontier on the brink. It's a part of the world that—like all relatively dark parts of the world—is in danger of being lost to light flooding. I am sensitive to the encroachment of light because I am watching it colonize the land I love most in real time. Yet here I am, with lights fixed on a fake-moon sheet, hanging UV flashlights from tree branches like ornaments on a Christmas tree.

I decide that I'll give the lights an hour or two to attract moths before I turn them off. Part of any successful moth-ing expedition is knowing when to quit. Part of any responsible use of artificial light at night, maybe, comes down to that.

Insect species are declining globally, with implications rippling through Earth's entire orchestra, creating a situation that the United Nations has identified as a food security crisis due to loss of pollinators. But studies have shown that even temporarily turning off lights during certain nocturnal periods, in cross-species compromises, can benefit some pollinators and migratory animals.

With moths, harm from artificial light doesn't come from collisions as much as from an inability to pull themselves away from light once they've been drawn. Light also increases their exposure to predators, and dealing with the shock of reaching a moon they were never meant to touch probably gets exhausting. Studies haven't definitively figured out how greatly artificial light is contributing to moth population declines, but there's solid evidence that artificial light affects both moth behavior and physiology, influencing their reproduction, development, and individual fitness.

Adult moths are famous for being drawn to light, but artificial light also affects them as caterpillars. With extreme exposure, caterpillars fail to enter pupal stages. In other words, caterpillars get so distracted by artificial light that their bodies forget to complete their metamorphosis. They cannot heed the ancestral call of natural darkness, which is needed for their transformation. Artificial light, in essence, steals their wings.

~~~~~~~~~~

In my yard, moth activity isn't immediate. Still, it's not a bad place to while away an evening. I can hear the river, happily gurgling. I can smell its silt, slightly sweet. I walk through a stand of pine trees as I wait, letting their needles brush against my skin like the bristles of a caterpillar's back.

I circle back to the sheet every so often. Each time, I find more dots. There are bits of glitter. There are mayflies. Just as

I'm about to give up on the arrival of any large-winged creatures, much less a giant leopard, I sense a fluttering.

A substantial moth has arrived, but the insect is moving too erratically for me to get a good look. The moth's tawny form crashes into the bedsheet, wings hitting cotton like the skin of a drum. When the moth reaches the sheet's edge, all movement begins to slow. Crashes turn to taps.

Sections of the moth's wings are orange sherbet, lovely even before I can make out any distinct markings. In time, the moth is completely still. It's as if this creature is waiting—frozen in some survival strategy.

This moth isn't as showy as a cecropia, but there is a line of black circles along the insect's abdomen. The middle dot is a perfect circle, but others seem to slightly bleed, making the moth appear to be tattooed with a moon-phase calendar.

I have an identification app loaded on my phone. In Ohio, surrounded by avid moth-ers who showed me not only how to access night, but also how to identify what I found in it, I didn't need phone apps. But tonight, I snap a photo for positive identification. When I get the results, I'm stunned: I set out in search of a leopard. Instead, I found a tiger.

This is an Isabella tiger moth, to be exact. But the real kicker, the part that leaves me in open-mouth awe standing in the damp grass of my own backyard, is that the moon-tattooed moth that has accepted an invitation to my Moth Ball is none other than a grown-up woolly worm.

I'm embarrassed to admit that, before tonight, I did not know what the most famous caterpillar of Southern Appalachia—maybe even North America, at large—became in its mature form. I had never followed the woolly worm's story past the fanfare of trademark festivals, since the larval form of this species, rather than the winged one, is what humans have chosen to laud.

Later, I will survey local friends to see if this is a personal oversight. Everyone I ask is familiar with the festival. But no one I know has ever really considered woolly worms past their famed caterpillar stage. Our neighbor is half-stranger. It's equal parts appalling and wondrous. Because instead of making me fret over the unknowns of this world—like so many of the harrowing unknowns of recent years—it gives me reason to think of how marvelous it can be to discover them. And here, in the middle of my own life, when it might have seemed as though new experiences and epiphanies were an endangered species, I'm only beginning to recognize it.

The artificial light that's drawn this moth is the same force that trapped my cecropia companion, leading to his ailments. The weaponization of light is a power I didn't know I had, a power I'm still not sure I want to accept. But it is undeniable. Artificial light doesn't just disrupt moths; it holds them hostage. To create a moth-ing station is like catching lightning bugs in a jar. Only here, light is the limitation.

Isabella tigers, like cecropias, live most of their lives in larval form. This winged beauty has only a brief window of life in flight. I have communed with woolly worms dating back to my earliest memories of the natural world. From the scraggly edges of my grandparents' farm to the loose stone of my gravel road, these famed caterpillars have been crawling the land I know best and tracing the lines of my palms with intimacy for decades. Clearly, Isabella tigers have also been encircling me.

Every time I use artificial light at night, I'm part of a

cross-species conversation. I call and call with headlights and flashlights and porch lights. But never have I paused to absorb moths' responses, not really. The more I learn about the world at night, the larger I understand my ignorance to be; the more I understand that night has been trying to communicate and I have not been receptive to its messaging. This moth has delivered the end—or, rather, the beginning—of a story I've long been part of but have only recently begun to claim.

There are, thankfully, things I can offer moth species, including this one, in the long term—native trees, undisturbed leaf litter, soil that is free of pesticides. But right now, just as darkness was the only solace I could offer that cecropia in palliative care, darkness is all this moth needs to thrive for however long she has left.

I hunger for more citrine wings. I have a compulsion to burn energy to see what might come if I keep my lures lit. But when I lean in to inspect the tiger's wings again, I flash to my cecropia companion's wings lifting at the wisp of my breath. Now that I'm aware of this Isabella tiger moth's existence, I need to facilitate a release. Natural night is not mine to colonize. It's time for me, as a human, to return what I've taken without asking.

Even when darkness is restored here, it will take time for this dazed creature to reacclimate. It will take time for me, too, to reorientate. How often it is that when we attempt to tame and claim wilderness—be it in the form of a mountainside with fences or an expanse of night with artificial light—we end up entrapping ourselves.

This tiger and I have been drawn together by artificial light, the proverbial flame. It holds us both captivated. It holds us both captive. And if this goes on for too long, it will be to our shared detriment. But I'm the one who built this box of light, and I'm the one who must act to alter it. Beam by beam, I dismantle our cage until the tiger and I are both free to roam this wider, wilder night.

Bats Flying

~~~~~~

Bat Blitz

Just a few weeks after Mothapalooza, I find myself in Alabama looking for a woman with a bat in her hair. It's the only directive I was given when I asked how I might recognize Vicky Smith, an environmental educator I connected with prior to traveling. She's my main contact here, at the Southeastern Bat Diversity Network's Bat Blitz event, during which biologists and volunteers from around the region will be surveying bat populations in Bankhead National Forest. As suggested during registration, I've come straight to the main office, which every attendee casually refers to as "Bat Headquarters."

The bat-in-her-hair description might not have caught me so off guard if not for the fact that, when packing for this excursion, I slipped no fewer than four hair ties into my luggage. Usually, I forget to pack even one. But somewhere in the back of my mind is the idea that bats are prone to getting entangled in long hair, and mine is overdue for a cut.

I'd like to think of myself as bat-tolerant, but I've arrived at the Bat Blitz looking to become downright bat-positive. I started paying attention to bats some time ago, when one took up residence in my porch eaves. I was wary about having a bat roost so close to my living quarters, but we went on about our respective business—with only a couple of dive-bombing incidents—until, one day, the bat died. When I found a carcass near my front door, I called a wildlife officer, fearing rabies. He told me to leave the bat alone until he could get to it. But before he arrived, a scavenging animal carried the bat away, potential pathogens and all. Even the memory is a little disturbing.

I've been trying to get to know these flying mammals, but it's hard to figure out how I might have an intimate encounter with bats without endangering all of us. So when I heard about this event taking place at the far southern end of the Appalachian chain, I decided to take a road trip.

Bat Headquarters is packed. There are friends reuniting with hugs and salutations. Sodas are fizzing. People who have traveled long distances are piling food onto plates as part of the welcome reception. Finally, I spot Vicky, who has a metallic bat tucked into her ponytail like a crown. She is wearing bat earrings and multiple bat rings that flash silver when she talks with her hands. Her shirt is covered in bat shapes. Her arms are stacked with bat bracelets. And word around camp is that she has a colony of bats back in her bunkhouse on the sprawling summer-camp property we've commandeered.

Here, Vicky is known as the "Bat Lady." She might be the only person in the state of Alabama to have a bat-carrying permit. They're hard to come by, in part due to the responsibility of caring for the animals and concerns about a variety of diseases, including rabies—something that I have been overconcerned

with, statistically. Roughly one percent of the wild bat population is thought to have rabies at any given time, and there are
generally only around two cases of human rabies a year in the
United States. Yet it is hard to have a conversation about bats
without someone bringing it up.

Vicky is caretaker of a small colony of rescue bats, which
she introduces at educational presentations. All her animals have
injuries that prevent them from returning to the wild, including
one bat who was maimed by a domestic cat, the most common
reason bats require rehabilitation. "We don't know a lot about
bats," she says. "If you see a raccoon and you want to learn
about it, you could probably follow it for a while, even though
it's nocturnal. But when a bat wants to leave, it's gone. It's hard
to get a good look at one in the wild."

This is one of the reasons that Bat Headquarters is so busy.
Events like this are one of the only ways that are considered
responsible close-range interactions with wild bat populations,
even for people working on bats' behalf the entire year. For several nights, the group will work to see how bats are doing—and
to better understand how land management plans might be
affecting them.

Wes Stone, a professor from Alabama A&M University who
is sitting with Vicky, has 25 years of bat research experience.
He's spent whole summers in forests, capturing and releasing the
animals to learn more about them. He was the first person to
find that white-nose syndrome, a fungal bat disease, had reached
the state of Alabama.

No matter how much care goes into figuring out how to set
up research traps, some bat chasers stay up all night without
seeing a single animal. It's always a disappointment, but it's also

a statement on the status of populations, which have, in many cases, been decimated. White-nose syndrome has caused a shocking 90 percent decline in certain bat species in just ten years. For bat chasers, finding bats has always been cool. But finding one of the most fungal-sensitive species now feels like a boon.

When I told friends I was headed to this event, many of their reactions gravitated toward fear. Bats, like darkness, are often associated with death. A psychology study found that they trail just behind maggots in negative perceptions. Sometimes, attendees at Vicky's lectures, including adults forced to chaperone schoolchildren on field trips, are so afraid of bats that they'll refuse to look at them. "If you're frightened by something," Vicky says, "it probably means you have something to learn from it."

When I bring up Mothapalooza—still riding high on my Isabella tiger encounter—one of Wes's students, Karmen, scrunches her nose. "Moths can do a lot to harm crops," she says. Bats are often defended as important since they eat insects at a rate that saves U.S. farmers pesticide costs and billions of dollars in annual losses. Bats also disperse seeds and, like moths, act as night shift workers, pollinating crops.

Catapulted into the unexpected role of moth public relations agent, I suggest that moths and bats, as maligned creatures, might benefit from their advocates sticking together. Karmen is dubious. But there are some things she appreciates about moths, even if she generally views them as pests. She says, "You're gonna love this: When bats go after moths, some of them talk back."

Isabella tiger moths, as it turns out, are particularly good bat communicators. When bats try to eat them, they make their own clicks, telling the bats that they should back off. Moths and bats have coevolved to a level that, when a bat is chasing a moth, a moth can tune its hearing to become more sensitive to bats' high-pitched calls. And all those moth wings I've been swooning over function as invisibility cloaks.

Moth scales absorb bats' ultrasonic acoustics. When a bat sends out ultrasound, moth scales resonate at frequencies almost perfectly matched, muffling echoes that might have otherwise revealed their location. Each scale on a moth's wing resonates at a slightly different frequency so that, altogether, they can absorb broadband.

I soon notice that conversations with bat people—which is how bat-chasers tend to refer to themselves—often gravitate toward counter talk about cross-species misunderstandings. "A lot of people seem surprised that bats have fur because the only bats they've ever seen are rubber ones used for decoration at Halloween," Wes says. "And people think they can't see, that bats are blind—but they see very well." Still, for bats, echolocation dominates. A bat can read a pine tree down to its individual needles just by how the animal's pulse-songs are returned to it.

Bats are generally divided into two categories: cave and forest dwellers. Wes's favorite species are the ones that like to snuggle under the bark of oak and hickory. There are more than 1,400 bat species in the world. Some are loners; others like to congregate. The smallest bat in the world is smaller than my thumb. The largest has a wingspan of three feet. "Most people are surprised by bat diversity," Wes says. "But there's joy to be found in it."

Unfortunately, bats' bloodsucking reputations prevent a lot of people from viewing them as anything close to joyful. Vampire associations are one of the most frustrating things Vicky deals with in attempting to better interspecies relations, though there isn't a single bloodthirsty bat species that lives in North America. "People think all bats want to bite them for their blood," Vicky says, shaking her head.

"Or that they're going to get tangled in hair," Karmen says. This draws especially hearty laughter from the group and, from me, a confession. When I admit that I've overpacked hair ties, I feel vulnerable in my ignorance, but I'm curious about how I absorbed the notion, and they seem like people who might help me figure it out. Maybe, I suggest, I'm just thinking about it because I've been dive-bombed so often? During my most recent neighborhood bat encounter, I didn't run back into my house, but I was rattled.

Karmen puts a hand on my arm. "You've got to remember, bats are better fliers than Tom Cruise in *Top Gun*. That bat wasn't looking to crash into you. He was just looking for supper."

The bat, she explains, wasn't after me. He was trying to catch the insects encircling my head. "Insects are gonna be attracted to you no matter what because they're attracted to the carbon dioxide of your breath. And bats are gonna be attracted to you because they're trying to eat those insects," she says. "You're basically a Golden Corral restaurant covered in snacks! You're a Sunday-after-sermon church buffet! To get insects and bats to leave you alone, you'd have to stop breathing!"

Given that I've been out at night more than usual, searching for moths and maggots and all sorts of things, I've been hearing bats. Hoping that these enthusiasts might be able to act as interpreters, I explain, "What I've been hearing isn't clicking, it's more like a tit-tattering." Before my direct observations, I hadn't realized bats—famous for vocalizations at pitches beyond my capacity to hear—made so many sounds audible to humans.

"There's a call they make when hunting," Wes says. "And one of my students thought she could tell when a bat was about to fly into a net. Sometimes they scream like they're saying: 'Help, help!' And the thing is, other bats do respond, like they're trying to help. It's something we see quite a bit."

Barely recovered from Mothapalooza, I can tell the late-night rigors of Bat Blitz are, for me, going to be a struggle, so I leave the reception to find my bunk. On my way out, a biologist with the same bunkhouse assignment stops me: "Just so you know, there might be bats in the lobby tomorrow morning."

"A presentation? What time?" I ask. I'm looking forward to seeing Vicky's bats, but I'd rather not be awakened by a crowd before breakfast.

"No, I mean the bats are staying at our place. The Bat Lady said she usually sleeps with them in her room but sometimes they keep her up. Since she's with bat people, she asked if they could just stay in the common area. They might already be in there, and I didn't want you to be startled. I mean, they're in enclosures, but still." Already, some attendees have started referring to our cabin as "the Bat House."

Unsure of what I'm going to find, I walk in complete darkness to the bunkhouse, where I discover that bats have, indeed, entered the building. On the ledge of a dormant fireplace, there is a line of towel-covered enclosures and a note, signed by Vicky: "Bats, do not disturb. If they disturb you, please come get me. They do squawk and fuss."

The largest enclosure is decorated with a button featuring a bat with glowing orange eyes and bared fangs. "Super Scary," it says. I've seen similar paraphernalia since I arrived at Bat Blitz, because almost all the relevant merch that bat people can find is related to Halloween. So they own the association, embracing holiday decor year-round.

The association of bats with Halloween traces back to the fact that, as they prepare for winter, bats are highly active around the holiday. It also has roots in Celtic Samhain, a festival where people would gather around bonfires for food and merriment. The firelight they created to warm themselves attracted insects. And to those insects, bats were drawn. The bats' emergence from darkness—followed by their fanciful flight patterns as they scooped insects for dinner—made the Celts view them as erratic spirits pulled from terra incognita.

Calling an eccentric person "batty" any time of the year might have origin in similar behaviors arising from the term "bats in the belfry." A belfry, as it turns out, is a bell tower. And, because—before electric lights—belfries were havens of cave-like darkness, bats liked to roost in them. When the tower bells rang, the animals were driven out and, disturbed, they often left in frenetic flight patterns. This made people think that the bats were inexplicably wacky when, really, they were just reacting to terrorization. Scared, they looked scary.

From underneath the towel that has been draped over the enclosure to shield bats from the bunkhouse's artificial light, I can hear them softly chirping. I came to Alabama hoping to commune with these elusive creatures in some meaningful way. Still, it's hard to believe that we're already roommates.

~~~~~~~~~~

Of all my bat housemates, the Egyptian fruit bat known as Aurora is by far the most handsome. I meet her the following morning at a safety meeting that's required for all bat chasers.

It includes presentations about venomous snakes in the region. There's also a session about downloading topographical maps, since where we're headed, there is no cell service and people are prone to getting lost. When the merits of local hospitals begin to be discussed, I start to wonder what, exactly, I've gotten myself into.

All the while Vicky's colony—which includes a trio of local microbats—is drawing a crowd. It's daylight, so their towel curtains have been removed. But, on schedule, they're sleeping. Aurora is much larger than the others. She has ink-dot eyes and wings that swoop around her body. Some fruit bat species are called "dog bats" because of how closely they resemble puppies. Unlike the microbats, Aurora is stirring, though not much.

The microbats—barely larger than the chubbiest of caterpillars—are most closely related to those we might find this evening. I know that if they spread their wings, they would take up a great deal more space, but the microbats look impossibly small when snoozing. Some bat biologists have noticed that people tend to report bats as being much larger when they see them at night as opposed to during the day. I suppose every animal expands or shrinks in relation to how greatly we fear it in the moment.

A staff member from the on-site kids' program has decided to stop by. The guy, who appears to be in his twenties, says without diverting his gaze: "I just can't stop staring. I'm used to seeing bats on TV shows—and that's exactly what I think of when I think of bats." He points to Aurora.

Before he started working at the camp, this counselor didn't know that bats lived in Alabama. He has, since getting a job outdoors, seen plenty of bats around the camp property. "Bats are so quick around here, I can't really tell what they look like," he says. "But I didn't expect them to look like this."

Even if meeting a bat colony can alleviate some of the physical fear of bats, there's still the worry of pathogens. For humanity, statistically, viral fear is the bigger boogeyman, especially in the wake of a pandemic. Thankfully, Bat Blitz is the kind of place where, just as you start contemplating things like this, you serendipitously end up walking a wooded trail with a microbiologist.

Kristina Burns, another of Wes's students, is at Bat Blitz because she never wants to be dependent on other people for her field samples. And chasing bats seems, to her, exceedingly fun, even though the schedule is challenging. "Once you have observed something in the wild and get tuned in to it, you're always going to remember. You're always going to be on the lookout. It's almost addictive."

Kristina wrote her dissertation on white-nose syndrome. But, since graduation, her work has branched into viral pathogens. There are, at Bat Blitz, several virologists in attendance. Bats have supercharged immune systems that allow them to carry pathogens that are dangerous to humans. The more we destroy their habitat, the closer to us they must come, and the more likely they are to transmit those pathogens.

Bats, one of the most populous mammals in the world, are the only ones known to fly, and it's thought that this characteristic might contribute to the superpowered immune systems that allow them to carry enormous pathogen loads, because it raises their body temperature in ways that might promote immune system activity—not unlike how humans fight pathogens with fevers. And bats' capacity to carry viruses might offer a form of hope that's rarely acknowledged. Because they play host to pathogens that, if let loose in the world, humans would not be able to absorb without falling severely ill.

Scientists might never be able to trace the exact origin of COVID-19, but it's been well publicized that some people have suggested it was transmitted from bats. Regardless of the specifics, the CDC reports that zoonoses, diseases that spread from animals to humans, represent a large percentage of emerging disease, and bats' already gloomy reputation has been further damaged by the pandemic, which inspired an outbreak of violence against them around the world. The U.S. Fish and Wildlife Service calls bats "one of the most important misunderstood animals."

The reality is, when we displace and kill bats, their pathogens are free to seek us as hosts. Still, we tend to frame bats as villains rather than superheroes who, under duress, are increasingly unable to maintain their status as living force fields that protect us by holding viral loads at a distance. We depend on bat health to hold back floods of all kinds of fatal-to-human pathogens.

In 2020, studies made advances in understanding bat behavior in Australia, where stressed fruit bat populations are leading to a growth of zoonotic diseases. These bats, which look not unlike my handsomest roommate, are known to immerse their entire heads in flowers, emerging with bright-yellow pollen on their puppy faces. It's a practice that plays a role in keeping forests diverse and resilient via pollination. Unfortunately, the loss of wild, flowering habitat has been pushing the fruit bats closer to human populations as they attempt to survive by scavenging on agricultural land—leading to contagion.

When a rare flowering event occurred in remaining patches of habitat, researchers found that it was enough to entice bats to return to wilderness areas, and human-bat disease spillover all but disappeared. The bats, getting what they needed while supporting ecosystems, gave greater distance to human settlements, leading to headlines like "Giving a bat flowers might preempt a pandemic."

We've characterized bats as moody goths, but they might be more akin to flower children. Still, we don't know them well enough to understand exactly what they need. It is what the Forest Service hopes to learn so they can enact updated land management plans. Unfortunately for humanity—which depends on reservoir species and remnant habitat to hold pathogens at bay—private landholders don't always take other species into account. And bat knowledge must be sought very carefully.

"Bats don't come sit on your knee like butterflies," Kristina says. "They have a beauty that's generally best observed from a distance."

~~~~~~~~~~

I've been told there's a survey site near a small cave. It's the only expedition labeled with a warning on the sign-up sheet: "Strenuous hike!!!" The exclamation marks are intimidating. I sign up anyway.

Every group has a local Forest Service guide, and my group—a collection of wildlife agents, interns, and students—has been assigned Ben Blair, a millennial wildland firefighter who grew up just a few miles from Bankhead National Forest. He is well over six feet tall and boasts a woolen lumberjack beard. In low-fire season, he has been reassigned to wildlife duty. Around Bat Headquarters, he is almost as well-known as Vicky, because Ben is a wildlife officer who is terrified of bats.

He is also afraid of frogs, given an encounter he doesn't want to recount. And don't get him started on ticks. He's not a fan of moths, either. But bats are absolutely one of his least favorite animals. Yet here he is, driving a Forest Service truck toward our survey site, a gorge that hardly ever sees humans. "I get it," I tell him. "I'm not a fan of spiders."

He agrees that spiders aren't great. Still, some of the only nocturnal creatures scarier than bats, to Ben, are snakes. "At the beginning of fire season, I stepped on a rattlesnake, and it didn't strike me. I've been around lots of snakes since then, and they haven't bitten. It doesn't mean they won't, but I'd never thought about a snake passing up a chance to get me. I guess maybe being around bats is going to make me think differently about them, too."

Ben has already met my roommate Aurora. But she wasn't much help. "When Vicky pulled those bats out at the meeting, I was so terrified I went into the bathroom to hide for a while." Ultimately, when the crowd thinned, he came out to look at them. "Those little ones with their wings tucked in weren't so bad, I guess. I was surprised about how small they are, teeny tiny."

"You were scared of the big one, the fruit bat?" I ask, amazed that he didn't swoon over Aurora's puppy-dog face, as I did.

"I was told bats were vampires from when I was little, and it really looked like one," Ben says, shaking his head. I also thought Aurora looked like a vampire, but in a lovable, Disney-character way. Movies and fairy tales of all types undoubtedly have a great deal to do with how we view animals, both beloved and seemingly scary. Psychological studies have found that children as young as five, kids who have never had direct encounters with bats, tend to categorize the animals as "bad," with researchers citing it as an example of how negative stories in media and culture work to shape natural-world fear from an early age.

Western mythos about vampires didn't originate with bats, despite common perception. Many associations sprang from rumors about a wealthy Romanian man in the 1400s torturing and murdering people. He was known as Dracula, and he was rumored to dip bread in the blood of his victims before eating it. Bram Stoker based his famous story on this Romanian, suggesting that he might be able to disappear in the night, as a bat, after committing violent acts.

It was human bloodthirst that led to the tale of Dracula. Yet, centuries after that mass-murderer died, bats are still paying the price. My housemate Aurora does not imbibe blood. She is currently, however, producing mother's milk. Unbeknownst to crowds, she has been sheltering a baby under her cape-like wings.

Ben's been thinking about Vicky's recent presentation—the part he saw before he started hiding. He was interested in what she had to say about how bats have gotten a bad rap based on how they're photographed, the only way most people ever see a bat face-to-face. Bat portraits are almost always taken while the animals are in distress. A bat that's getting a close-up is generally one that's being restrained under artificial light. Yet when people see the resulting images, they criticize bats, as if to suggest that they should smile more.

"Gah, in real life bats are such little chicken nuggets. They're just so cute! I love them!" one of the volunteers exclaims. "They can look a little scary when they're caught, though. That's true. It's just like in the pictures. Their mouths are open like they're screaming, but you can't hear anything."

Transcribing the silent screams of bats is part of her day job as an audio technician who conducts surveys. To do so, she often rides around nights with microphones on top of a vehicle to record the echolocation pulses we, as humans, cannot hear. Then she runs them through equipment that turns them into squiggly lines on a screen, visual translations.

In general, she can tell species apart by how many pulses are recorded and the patterns they make. Some calls are transcribed into rocky lines on her screen. Some look flat. "When pulses come close together, like static, it seems like that's when they're scared of us, but we're not quite sure yet." Regardless, it is a manifestation of bats' silent song. She has found a way to turn inaudible language into shapes that human minds can process like sheet music.

The cave-adjacent site we're traveling toward is a full hour's drive from Bat Headquarters. Halfway there, conversation lulls. Ben turns the radio up. Outside, scenery flashes in a Southern backroad Morse code: *House. House. Church. House. Church. Church.* Nearly every building is flanked by a security light.

"What's the coolest bat you've seen up close?" an intern asks, passing time. "Mine is probably an evening bat."

"Red bat, definitely," says the technician.

"I've seen red bats before," Ben says. "Not close up—from a distance, during a burn."

Until just a few years ago, scientists could not figure out where red bats, which tend to roost in trees, went during winter. They seemed to just disappear without a trace. Recently, it was discovered that those bats—colored to blend with the trees that shelter them—track leaf season, falling to the ground with leaves in autumn, making cozy insulated leaf-caves for themselves.

It was a shock to bat researchers that the animals weren't

changing location, just strategy. They were simply tracking the life cycle of trees. It's the type of discovery that Bat Blitz might provide. A change of perspective always has potential to shake loose new knowledge.

There must have been a hundred red bats rising from leaf litter when Ben saw them in that fire. He remembers their dark bodies above white smoke, all of them chased by flames that were licking oxygen from the air so that they, too, would have the strength to keep moving. "It's interesting, the way a fire behaves," Ben says. "It's like an animal in its own way."

The rest of the Bat Blitz volunteers are envious that he got to bear witness, though the circumstances, in terms of habitat disturbance, were upsetting—and illustrative of why it's important to understand bat behavior to modify land management plans. Still, for most of the bat lovers, it would have been bucket-list-wondrous to see a cauldron rising like that. For Ben, it was among the most horrifying things he's ever witnessed.

He turns onto a rutted dirt road that's guarded by a metal gate. "When I first saw them, I thought they were birds taking off," he says. "I would have seriously freaked out if I'd known they were bats at the time. I'm glad I didn't find out until they were already gone."

Close Encounters

There is no trail where we're headed. No one knows what bat species, if any, we'll find tonight. Some of the bats at our assigned site might be federally endangered due to white-nose syndrome, and others might be eligible for listing if their population numbers have greatly gone down. This, in addition to land management plans, is one of the survey's main functions.

It is not actually the fungus that grows around the noses of bats that kills them; it's that it disturbs their winter rest, forcing them to be active in the wrong season, when there are no insects for them to catch. They exhaust themselves trying to get what they need in a world where they cannot find alignment. In the restlessness of their discomfort, they expend all their bodily energy and, unable to replenish it because they're out of step with the seasonal dance of their food sources, they ultimately perish.

Halfway into the ravine, someone pauses, creating a domino-effect up the mountain. "What's going on?" someone shouts from the back.

"We're just appreciating," comes a message from the front.

Golden light is filtering through trees. In a group of more than a dozen people, nobody questions the validity of a sunset-appreciation pause, even though we're nearly sitting on soil due to the severity of the ravine's incline. Previously, we took a break to pass a giant snail shell down the line. It's a lovely pace at which to hike.

We left our Forest Service vehicles on an access road above. Even with four-wheel drive, the trucks had been slipping. There is a small chance that game hunters have seen this place, but probably not for a while. Alabama isn't often lauded as an ecological treasure, but in 2020, it was shown to be one of the most biodiverse states in the nation. Bankhead is part of an area that's occasionally called America's Amazon. It's rainier than Seattle and thought to have more oak tree diversity than Great Smoky Mountains National Park.

At the base of the mountain, our crew flows directly into a dry riverbed. It's wide and doesn't seem to have a puddle in it, even though we drove through storms to get here. Even now, rain is thumping against leaves, though it cannot touch us. The overstory is so thick we can no longer see sky.

I balance on large stones that have been polished by long-gone waters. When we take a break to wait for stragglers, Pete Pattavina, a bat biologist with the U.S. Fish and Wildlife Service, points out soft beds of greenery that are growing along the edges of the rocky path, identifying them as dwarf crested iris and liverleaf. "They're ephemeral wildflowers. Imagine these all blooming in spring." What seems a deadened pathway of stone is, in season, a river of flowers blooming blue and violet, white and pink.

Ben and a group of the others move downstream, toward an area where there's flowing water. The ravine is too deep to radio out with walkie-talkies, so we've left a sentry at the trucks to ping signals. When we reach a halfway point between our encampment and the one Ben's stationed at, Pete decides we should erect our bat-catching mist net. To catch bats, Pete explains, it's important to identify flyways—runways in the sky. "We're like state troopers setting up a speed trap on a busy road," Pete says. "We want to funnel animals into an area."

He pulls out a piece of bent rebar meant to hold our mist-net poles. "Hard to imagine what could have done this," he says. "Somebody must have had a bad day."

However it happened, the off-kilter equipment stands to make our job harder. Other members of the group have scattered. No one on-site knows how to proceed. We're all short, and the pieced-together poles we're dealing with are over twelve feet in length.

Our goal is to erect them in the forest and run a thin-filament net to catch bats in midair. The trouble is, the ground is stone, so we cannot easily drive the poles in. There's soft soil on the edges of the river, but it's hard for us to even hold the bars erect. They're too tall and wobbly.

I gather rocks to help anchor them. If we don't get the mist net in place by the time bats start to become active, we risk this being for nothing. "We don't have much time," Pete says, dismayed at the state of the net, a jumbled mess.

"I believe in us!" cheers an intern.

"Hey, I'd rather be out here than on a computer, no matter how long this takes," Pete says.

I walk up and down the substantial net, trying to tease out the knots, as the others attempt to hold the poles in place. All the while, Pete is shouting for us to check for copperheads. Many snakes are nocturnal, same as spiders. And if we're on the verge of being too late to get out of the bats' way, we're also on the cusp of meeting their nighttime compadres.

Our chances of finishing before dark are not looking good. Then, from downriver, reinforcements. Someone has gone to get Ben. "I heard you could use help from some tall people," he says.

Pete is visibly relieved. "What's your name again?

"Ben. People keep telling me I'm easy to remember because I'm the guy who's afraid of bats."

"You're afraid of bats and you're out here—that's awesome!" the intern says.

Ben shrugs and takes the towering rebar from Pete, who tells him that there's no mallet in the bag. This doesn't give Ben pause. He picks up a rock and begins to hammer, stone against metal. "Afraid of bats, my ass!" Pete exclaims in admiration of the muscle flex.

"Still afraid of bats," Ben says. Fear, after all, isn't something that can be strong-armed into submission; it's a complex entanglement.

Once the bars are up, the net unfurls as dark gossamer and Pete runs it up the pole like a flag. Ben is responsible for relaying distress calls, so he heads back downriver.

With the mist net secure, I make my way upstream, preoccupied with avoiding snakes. When I see the audio technician crouched in the riverbed along the way, I ask if she's found a snake. She has a headlamp on, appearing as a Cyclops, with one shining eye. "No snakes," she says. "I'm looking for little sparkles. That's what I call them."

Her little sparkles are, as it turns out, spiders. She knows a trick to appreciating them. Spiders—like many other creatures with eyeshine capacities, including dogs and cats and frogs—have an iridescent layer of tapetum that reflects. At her urging, I adjust the light that's been hanging around my neck. I think about how, in certain seasons, this is a river of flowers. And right now, it is glistening with the eyes of spiders.

Spiders are, like bats and moths, crucial nocturnal pollinators. They have a special relationship to flowers and—in alliance with bats—play a great role in securing human food crops. Without spider and bat participation in agricultural activities, humanity would lose vast food supplies. Even after I learned this, though, the news didn't do much to change my perception of spiders. But what I'm seeing here is that life-sustaining importance translated into beauty that might be powerful enough to override my low-grade arachnophobia.

The riverbed is overflowing with eyes that appear as iridescent pebbles in headlamp light. When I move my head to the left and right, their twinkling covers the riverbed and spills into the forest around us—light reflected by the dewdrop eyes of what must be more than a thousand spiders.

I focus on a single pinprick orb and crouch to zoom in, soon wishing that I hadn't. I find a spider the size of a small tarantula staring back. Then I zoom in on another spider, smaller than the tip of my pinkie. There are countless species, and the prehistoric vibe of this forest makes it feel like there might well be creatures not yet recorded by science here.

I take one step back. Then another. But the spiders have me surrounded.

From the angle of my headlamp, I can see them in every direction. I could, at will, pinpoint and go after a spider that's nearly a football field away. Only I don't want to. That's not surprising. What's surprising is the realization that the glittering currents aren't moving toward me, either. We're at a standstill. I don't want to harm these spiders, and they apparently have no interest in pulling me asunder. In any case, we, the humans, are the ones who've erected a giant, ensnaring web this evening.

My light, reflected, seems like confirmation that these spiders can see me as I now see them. But neither of us is making a malicious move. This riverbed is the path of least resistance through this gorge—by air or by land—which means it's a throughway that plenty of animals, seen and unseen, are sharing. The phenomenon of spider-shine isn't one that I'd want to observe often at close range. But, from a distance, I cannot deny that this waterway of opalescent eyes is teetering on miraculous beauty.

I try to shake off my spider bias. I do my best to accept the scene for what it is: gorgeous. Given the shimmering Milky Way that's winding along this dark forest floor, I mostly succeed, as I hop stone-to-stone across the newly discovered galaxy.

~~~~~~~~~~

Pete suggests that I should walk upslope to check out our bat cave of interest before the animals become active. Its entrance is a small one, barely large enough for a couple of humans to slide through comfortably. It's shrouded by a harp trap that others have set up while we were working on the mist net. The harp trap is, amazingly, exactly what it sounds like: a version of the classical instrument, only, instead of sitting on an opera house stage, it's guarding a bat cave in Appalachia.

Historically, these traps used metal harp strings. Now, researchers use clear filament that's a bit more forgiving. There is no way around the strings, so a bat coming out of the harp-blocked cave is required to navigate. Generally, echolocation helps the bat navigate between the first set of strings. But the double-strung harp has a second line of strings that bats cannot maneuver in quick succession, causing them to fall against cords that guide them downward into a canvas bag.

At a makeshift bat-processing station below the cave—a metal camp table that has been laboriously lugged into the gorge—I join Pete and Nick Sharp, a state agent from Alabama Wildlife, along with the rest of the crew, to sit cross-legged on the forest floor.

Idle minutes pass without a bat. Then, an hour.

"I don't have a good feeling," Pete says.

Nick shrugs. "Maybe they're not active because of the rain." The muggy heat has turned into a damp cool. The whole ravine feels like the interior of a cave.

In time, headlamps get turned off. Our restlessness disintegrates. After a period of listening to zipping backpacks and the crunching of packed snacks, Pete's voice pipes up: "There's something glowing over here."

He's found a railroad worm, a millipede-looking larval beetle named for patterns that resemble lit windows of a passenger train. Segments of the railroad worm's body are glowing brightly. I have, in the span of a couple of hours, seen almost as many millipedes here as spiders. Maybe it's because this place is particularly rich with them—or maybe it's just because millipedes are nocturnal, active when I'm usually not, crawling on bare ground, where I'm typically not sitting.

Pete lifts the railroad worm. The creature curls into a small ball, a fixed point of light in the center of his palm. "Good job, Pete," someone says.

He laughs. "It was effortless. All I did was look down."

We're still reveling in the discovery when there's a shout from the direction of the harp trap: "We've got a bat! Tricolor!"

"Are those rare?" I ask.

"They are now," Pete says. "Tricolors might skip 'threatened' and go straight to 'endangered.'"

We prepare the table, laden with magnifying glasses and scales. Pete, suddenly stern, shouts: "Masks on!"

The medical-grade masks we've been given are not so much intended to protect us from contracting diseases so much as they're meant to protect the bat from reverse spillover as COVID variants linger. Humans and bats are not meant to be as close as this, and our masks acknowledge it.

Even those of us who are keeping our hands to ourselves are responsible for wearing respiratory gear and keeping track of equipment that will, ultimately, need to be decontaminated due to potential white-nose spores. I adjust my mask's double straps. I pinch its metal nose clip. Of all my pandemic worries, this is the weirdest: I've got to be careful, or I'll potentially expose a bat to my pathogens.

The bat has been carried from the harp in a brown lunch bag, which is placed in a plastic cup so that the animal can be weighed. The paper is trembling. And, from within it, the bat is click-clicking.

"We have to make sure we can get the transmitter on this one," Pete says, pulling out a small board that I recognize as part of a transmitter kit from Copperhead Consulting, a firm that's best known for its work creating faux bark that can be wrapped around structures to mimic natural habitat. If this had been nearly any other species, we would not be attaching a transmitter. But tricolor bats have proven particularly sensitive to white-nose syndrome and are of special interest. Finding this one is a hopeful sign of populations hanging on.

Pete unpacks a small soldering kit. For a minute, it doesn't heat. But, then, a small spot of orange. He solders a small wire to the device that will activate the transmitter. It will be attached to this bat like a high-tech version of the stickers put on monarch butterflies by citizen scientists.

"Is the receiver on?" Pete asks a twentysomething volunteer, who nods. Pete has told me that it's hard to get into this work, and he views the Blitz as a classroom. Pete narrates everything he does. He and Nick are the elder researchers in the group. Staying up all night is, as Pete's told me several times, mainly a young person's game. It won't be long before he's ready to retire from field duty.

"You want to make sure you don't overheat the transmitter,"

he says. I take a few steps back when he opens the bag to retrieve the bat. An intern notices me jump. I shrug. I'm a little nervous to be at close range, but I trust that these biologists know what they're doing.

The bat has been chirping, but when the animal is taken in hand, she goes silent, though her mouth is gaping. It is the exact open-mouth expression of almost every bat I've ever seen in a photograph. In still photography, this expression makes bats look like they're trying to bite something—or someone. But, here, even though I cannot hear a thing, I can sense a silent scream. This is not an expression of aggression; it's an attempt to communicate. It is likely a message akin to: *Help me.*

"Female, post-lactating," Nick reports.

This bat has already had her pup. Bats only give birth to one or two a year, and they've been known to be able to locate their offspring by voice alone in roosts of thousands. Nick slides on a metal band that will help future bat chasers identify her, making sure to orient the band so that it can be read upside down, since that's the roosting position in which many animals are found.

Pete uses tiny scissors to trim fur so he can apply adhesive. "She's not happy with you," someone says.

"No, she's not," Pete acknowledges. "She's fussy. And when she tells her friends she's been abducted like this, they're probably not going to believe it."

A small moth lands next to the bat, nearly slapping her face with powdery wings. It looks like the bat is calling the moth with a siren song, though the moth has likely been attracted to our lights. It is a peculiar sensation to recognize that we cannot perceive what is going on right in front of us, like watching life with the volume off.

The glue Pete's using is a medical-grade adhesive that will, in time, wear off. "You only get one chance to do this," he says, taking a deep breath.

The transmitter is the size of a rice grain. Pete holds the bat with one hand. "They have an indentation at the back of their heads. It's a natural place to put your finger, and it immobilizes the bat so that they cannot reach your hand to bite it," he says.

Once the transmitter has been attached to the bat's skin, Nick has the honor of releasing the animal, which is an art form unto itself.

"If you toss a bat into the air, you might think you're giving the animal a head start, but that might cause the bat to plop onto the ground," he explains. "They might be chilly and need to warm up. They might not be ready to go. Sometimes, they just want to hang out in your hand for a while. Every bat is different. We need to let them go under their own power."

Nick's hand blooms finger by finger, until it's fully open. This bat isn't interested in loitering. She's ready to take this transmitter on the lam. Since bats don't have the same lift as bird wings, to gain momentum, many need to fall before they can fly. The animal draws a blurry checkmark in the air, and we remain silent as the tiny bat we've examined stretches into a larger creature that, almost immediately, vanishes in the dark.

~~~~~~~~~~

Later, Pete holds up a receiver that looks like an old television antenna to see if he can determine where the bat went. "That tricolor is up the hill somewhere," he reports, machine beeping. "She's probably on a slope trying to chew off that transmitter!"

Now that the bat has been affixed with a tracker, Copperhead Consulting representatives will attempt to find the animal so that scientists will be able to fill in some informational blanks about habitat. "You can locate the roost sometimes. You can learn what habitat they prefer," Pete says. "Sometimes, all they require is a cluster of leaves. You get a feel for species. You start to understand what we're doing that's helping and hurting them."

He pauses. Insect song fills the air, so loud it makes my eardrums buzz. "I thought I just saw a bat fly down," Pete says. "But it could have been a giant moth."

The whirling tails of luna moths, maybe the most beloved of all moth species, evolved because of bats. Those streamers are not just for beauty's sake; they act to scatter bats' echolocation—keeping the moths from becoming prey. Details like this are everywhere, examples of how we were all born to dance together. And, continually, I find myself among people who've dedicated their lives to better understanding humans' role in the arrangement.

From the direction of the net, light seems to have multiplied. Two volunteers from Ben's group have walked up for a visit. They're passing the mist net along their route. We watch, hopeful.

"They have a bat!" Pete says.

"Let's take bets," someone says. "I bet they have a red bat."

"Big brown bat."

"Indiana bat."

"You'll get points if it's an Indiana," Pete says. "That's pretty unlikely."

"I bet it's a tricolor bat with a transmitter on it," Nick says. The group groans.

It turns out to be a red bat, fluffy as a miniature teddy bear. "Look at that fuzz!"

"Oh, beautiful!"

"I just love red bats."

"They're sensitive," Nick says.

"They're powerful," Pete comments.

Someone has brought a long, flat light. They illuminate the bat's wings from behind. Nick pulls out a small magnifying glass. "Look at those joints," he says. "They're not completely formed. This is a juvenile. Adults calcify; that's how you can tell age."

This bat isn't audibly fussing. This bat, from the beginning, is all silent scream. Red bats, forest species who roost in places where air circulates, fare better than cave species with white-nose rates. This meeting is just a record—there will be no trackers attached.

The two volunteers who delivered the red bat from the mist net are amazed that they walked by at just the right time to see her wings catch. "It's like she got so distracted by our lights that she flew right into the net!" one of them says.

Bat Conservation International has found that bat encounters often take place on patios with artificial lighting or in swimming pools where bats swoop toward insects that are swarming, often in ways that make people view bats' erratic flight patterns as upsetting. Bats are often found near streetlamps because they are looking for insects, which are drawn to illumination. But bats' eyes are adapted for low-light conditions, so they're often painfully dazzled by bright lights. Studies have shown that their hunting accuracy is lowered in the presence of artificial light. Without natural darkness, they cannot access the best of their abilities.

~~~~~~~~~

One brown bat and several hours later, we emerge from the ravine. The hike out was strenuous, as advertised. It was also discombobulating. We were able to find our way back only thanks to a trail of glow sticks that some night-familiar biologist had scattered on our way down, reminiscent of Hansel and Gretel's breadcrumbs. As we were climbing, I found a wild turkey feather, which, used as a tool to swipe the night, saved me from being ensnared in spiderwebs, a fate that befell many of my companions.

Already, we're late for Bat Blitz curfew, and now the rigors of decontamination are slowing us down. Bat fieldwork requires a level of biohazard protocol that looks like a scene straight out of a documentary about the CDC. Anything that was touched requires being wiped down with disinfectant. Post-chase camaraderie mainly consists of passing spray bottles to ensure that, on shoe soles and backpacks, any white-nose fungus encountered will not travel with us. Clothes are shed and put into plastic bags. Masks are disposed of safely.

"How was your introduction to the bat world?" Pete asks when he sees Ben.

"Well, one bit me and I started to develop superhuman powers . . . just kidding!" Ben laughs. "I stayed away from the first bat. Actually, I ran across the riverbed to get away. But the other two we caught, I got closer. I don't know. It's interesting— they didn't seem all that scary in person, not out here."

Ben's surprised that his fear has not expanded in proportion to bats' widespread wings. The animals seemed bigger in the wild. But his courage seemed to swell in the open air, too.

"Would you say you actually like bats now?" I ask as we get into his truck, smelling of disinfectant.

"I still think their faces are scary," he says. "But, I have to say, I like their little feet. Bat feet are really, really cool. I might even go so far as to call them cute."

Flight for Life

I sleep until I cannot sleep anymore. It is not enough. When I make my way into daylight just a few hours later, I find the entire community of bat chasers is rousing. The tricolor bat that my group found was the only transmitter-applied bat of the night. But one animal was all it took to activate Copperhead Consulting protocol.

Antennae-carrying technicians are already navigating the steep terrain of the ravine, attempting to catch up with the bat. Now a flight crew has been called in, led by pilot Steve Samoray, a lanky guy with shaggy gray hair that makes him seem more surfer than snake handler.

I run into Steve as he makes his way out of camp, headed for his plane, an actual Bat Mobile with antennae attached to both of its wings. When he takes off from a rural airstrip, he won't have a destination. He'll be wholly dedicated to searching for signals emitted from the transmitter on that single tagged bat.

"She could have traveled a mile or two by now," he says.

"Sometimes you'll think you've figured out where they are, and the equipment tells you to go farther. Turn right, turn left—when you're flying, you just do whatever you have to do to figure out where they're going."

No matter what technology you're using or how many miles you travel, it can be hard to find a bat in the mountains, where topography makes signals bounce in ways that confuse even the most seasoned technicians. They are searching for clues that might help landscape management plans promote life. Sometimes, a single tree left standing can harbor up to 400 bats. And that same tree, taken down, can leave 400 bats homeless, searching for other places to live, potentially closer to human development. So locating a single roost tree stands to make a great deal of difference across species.

When Steve learns that I've been staying in the Bat House, he reveals that, though he's worked with bats for well over twenty years, he's had some revelations at this event. "I've encountered thousands of bats in the wild, but when I saw Vicky's bats, I realized I'd only seen a handful of bats in sunlight," he says. "Sometimes, people will send me pictures of a bat hanging from a tree in their yard during the day, and I don't know how to identify it. In daylight, the colors are so different. Everything looks more detailed, more vibrant."

If Steve and his crew pick up on a bat signal today, there's a chance that the ground crew will have to follow up by knocking on doors, carrying antennae for bat chasing, trying to convince private landowners to let them poke around on their property. "Bat biologists have it easy," he says. "People can get behind bats."

"Really?" I say. "Pretty much everybody here has told me

that their friends have a hard time understanding why they even want to think about bats."

Steve shrugs. "When you go up to a house and tell them you'd like to look around their property when you're tracking bats, they seem pretty accepting. But when you tell them you want to look around their house for snakes, they're like, 'What are you going to do with that?' At least bats have legs. I think fear of snakes traces back to the Bible—serpents and all that. There's always potential for something to happen with wildlife. With anything. To appreciate bats, you learn to keep your distance. But you also learn that you don't have to throw your hands over your face. You don't have to cower or cringe. Most of the time, when you encounter a bat, all you need to do is chill and it'll fly right past."

~~~~~~~~~

Weeks later, Steve makes good on a promise to follow up with a tricolor bat status report. The Copperhead crew was able to pick up on the bat's transmitter signals after we parted ways, but they soon lost her. No roost was ever located, but not for lack of trying.

Steve hunted for that tricolor bat by plane for days, flying search patterns across Bankhead as, on the ground, teams of technicians hiked through forest, some of them even descending into the tiny cave that had previously been covered by the harp trap. Still, nothing.

"Likely, she found another nice cave to hang out in during the day where our receivers wouldn't pick her up," Steve mused, "or she flew out of the forest and is living in someone's backyard."

To think that this wild bat, important enough to convene what seems like a military-grade operation, has ended up hanging out in someone's backyard in suburban Birmingham is almost unbelievable. Basically, if you see a bat, it is best to assume that the animal is something special.

Tricolor bats don't fly much beyond a 50-mile radius, even in seasonal migration, so I have no chance of seeing this particular bat in my yard, though we're connected by a long mountain corridor. But there are plenty of other bat species around. I even have a small cave in my neighborhood, and there are trees for roosting on surrounding acreage that has, so far, escaped development.

At Bat Blitz, it seemed a novelty to be rooming with bats, but in reality these creatures reside in backyards and parks all over the place. In almost every form of habitat. In all fifty states. On every continent except Antarctica. Before Bat Blitz, I'd never really been able to appreciate a bat encounter for fear of what might come next.

After all I've learned, I'd be happy to host a bat under my house's eaves. But I understand that my previous porch bat's activity was partially a message that, in an increasingly developed world, my winged neighbors might need help finding alternative arrangements—even with remnant habitat around. Brushing against moths might be okay, but the bat-human relationship is—like a well-performed tango—one that's designed to hold tension in distance.

The Leibniz Institute for Zoo and Wildlife Research has found that planting trees in urban areas can lessen human-bat interactions while protecting bats from the ill effects of artificial light. And when people erect bat houses, it gives bats places to roost away from porches as bats work to manage insects like mosquitos and the potential pathogens they themselves carry.

There are ways for humanity to dance more gracefully with bats, and that choreography can start with each of us—as individuals, wherever we live. If we can find ways to protect darkness and the various, oft-misunderstood life-forms it holds, darkness might, in turn, help protect us.

I resolve to pick up a bat box at a general store across town. They are typically slender versions of bird boxes that mimic the comfort bats find when they slide under tree bark. I also decide to plant some deciduous trees, since I've met so many moths and bats who depend on them. I'm glad to know that Copperhead's bat-bark technology exists. Still, I'd like to do what I can to prevent the making of a world where faux bark is needed as life support. But before I have time to make any of these yard alterations—enacting my own landscape-management plan—a neighborhood flier instigates a meeting.

I'm standing in my driveway when I see the bat swerving toward me. I don't hear a peep, though the way she is moving— in a topsy-turvy pattern, not a straight trajectory—indicates that she's hunting. This might be a brown bat, a common species in my area, or a tricolor bat, since white-nose survivors still fly here. I don't want to touch this animal any more than I wanted to be a designated bat-handler in Alabama. What I wish to learn is how to, at a distance, be a better-attuned neighbor—starting now.

When the bat flies toward me, I don't cower. I stand at attention, witnessing. Since I have seen a bat silent screaming from the confines of gloved hands, I wholeheartedly believe this animal wants to keep her distance as much as I want to keep mine. Why, then, have so many storylines pitted us against each other?

After Bat Blitz—where I kept my hair secured in a bun—I

tried to figure out where I might have picked up the hair-tangled-bat concept. Hair-related folklore about bats appears around the world, in a variety of cultures, with countless iterations that often involve bats as shape-shifting witches. Groups of bats in flight are, after all, known as cauldrons. But many people believe that in Western folklore, tales of hair-tangled bats have historically been used as a way of controlling women's mobility, warning them against leaving their houses after dark. While darkness can be full of danger, bats have long been taking the fall for crimes they haven't committed. And bat positivity, as it turns out, might just be an unexpected form of smashing the patriarchy.

This bat and I are taking back the night right here in my driveway. Frankly, I remain a little uneasy about the situation—and that's okay. There's a function of fear that's useful. It acts as a reminder to give space. With bats, as with lots of things, there is safety in distance. I wouldn't want to get so comfortable with bats that, at close range, I had the urge to pet them. But fear, taken too far, has the potential to make any animal act erratically. So how might I inch away from a fear of bats while maintaining appropriate distance? I've come to think that what bats need—along with many nocturnal beings, and even night itself—might be a renaissance of reverence.

Reverence—an intermingling of love and respect—is a positive, awe-leaning emotion that still encourages situational distance. It's often associated with religion, because—like those Celtic festivals full of bats and bonfires—it requires that we defer to the concept that there are greater forces at play, things we do not fully understand. The ecology of night, for instance.

In the oasis of my unlit driveway, there are no point-source lights to distract. There are no flyway traps. This bat has darkness to roam. This is a *Top Gun* flier working in decent conditions, and in her abilities, I have faith.

She can see me, not only with her eyes, but as part of the musical map she is drawing in her mind. This bat is bouncing pulses off the contours of my body, maybe even the lines of my face. There might be, in the swarming insects around me, a clicking Isabella tiger moth in the interspecies conversation taking place. I myself am not a passive bystander. Bats might be associated with death, but I summoned this animal with the carbon dioxide of my exhales, which means that every swoop of her dark wings is a welcome sign that I am breathing. And, post Bat Blitz, my hair is no longer cinched. In fact, it's loose and flying wild this evening.

Foxfire Glowing

~~~~~~

Vision Quest

Tal Galton has been walking backward for a quarter mile. Over broken stone. Over fallen leaves. I am among a handful of people following him, none of us bringing up his peculiar choice to not look where he's going. Walking backward is counterintuitive. But so is walking in the dark. Yet here we are, traveling toward dusk, because we are on a quest to glimpse foxfire—glowing fungi known to turn forest floors into scenes from a fever dream.

We all just met as strangers, with barely enough daylight for us to make out faces. It seems like Tal might be walking backward to use every ounce of sunlight left to study our identifying features. His stance also has the effect of assuring us all, in a language without words: If I can walk this trail backward, you're going to be able to walk it in darkness.

Tal takes a step back. We take a step forward. Anything seems possible. We're already part of a carnival parade.

Gravel turns to stone, polished from years of human travel.

"The path we'll be walking in full dark is similar to this," Tal says, helping us chart a way forward. "It'll still be an hour before it starts to get completely dark. By then, you'll hopefully have a sense of the surfaces we're dealing with. Tonight, we need complete darkness. We can't use any flashlights. The foxfire we're looking for is quite dim."

I first heard the term "foxfire" as a child, when my mother shared her copies of the *Foxfire* book series, which was founded in the 1960s as part of an ongoing oral history project that is still being updated today. The books have, for generations, recounted the stories, cultural practices, and land-based traditions of Southern Appalachia with large audiences. With nine million copies in print, its publisher calls the series "an American institution." But from my perspective, the books simply covered my family's everyday practices—things that, as a child in the faster-flashier 1980s, I mostly took for granted.

Even though I grew up with the *Foxfire* series—among grandparents who embodied the self-sufficient folklife in its pages—it was only recently that I learned foxfire is not a particular species, as I've long thought; it's a blanket term for bioluminescent fungi. Only 80 of the roughly 100,000 species in the fungi kingdom are known to glow, not only in Appalachia but across the Americas, Asia, Australia, and Africa. Despite this global presence, foxfire remains one of the most readily identifiable symbols of Appalachia, thanks to the book series. For me, finding glowing fungi has become a priority.

The purpose of fungi's bioluminescent function remains unclear, but some species use it to attract insects that might help spread its spores. Prior to electricity, people harvested foxfire to read by its light at night. Early submarines mounted it behind glass. Its chemical compounds are used to track things in the human body like infections and cancer cells. Historically, logs alight with foxfire were arranged on the ground to outline nocturnal paths. In 2015, a team at the Institute of Bioorganic Chemistry found a way to isolate a fungal protein to create luminescent plants that they suggested might one day serve as energy-efficient streetlamps, though green light emitted from all directions would present a light pollution issue. It all sounds terribly newfangled. But in some ways, it's the kind of creative, living alternative to artificial light that my Granny and Papaw might have appreciated.

"You never know what you're going to find on any given night out here," Tal says. We might find mycelium and bitter-oyster mushrooms. There could even be a jack-o'-lantern somewhere in the vicinity, with fruiting bodies that can grow as large and orange as their namesakes and are often mistaken for chanterelles, a foraging delicacy. "Eating jack-o'-lanterns won't kill you, but it might make you wish you were dead," he says.

One of the wonders he seems most excited to show us is a species he knows where to find: a minuscule mushroom that, as far as he can tell, is currently undescribed by science. "A lot of people who know a lot more than me have told me that they've never seen anything like it in this region," he says. He's currently working with mycologists to investigate. But, regardless of its scientific status, we're assured that we'll be seeing something that relatively few humans have ever beheld.

"Every so often you'll hear about a new population of something discovered on Grandfather Mountain or in a national park. But diverse fireflies, or lightning bugs, are everywhere around here," Tal says. "It's the same with a lot of foxfire species. To find them, all you need to do is go into the woods and really start looking." Only, to do this, some people—including me—need lessons in how to navigate the world nocturnally. Tal is one of the few naturalists I know who specializes in the practice.

The road we're traveling shifts from stone to dirt. There's plenty to see, and some of it is probably glowing, but the lingering light of day prevents foxfire—which glows constantly—from being spotted. So we walk as we wait, surrounded by what we're seeking, unable to see it. But Tal assures us that our bodies are already working to assist in our quest. "Your eyes," he says, "have already started to adjust for night vision."

The mention of night vision seems to create a subconscious panic. Group members start looking around, attempting to record everything possible before daylight leaves completely. But, despite this impulse, we're not here to escape darkness. We've traveled to Tal's home turf in Celo, an intentional community that has preserved its forests, because we want to find it here, along with foxfire.

I've visited Celo throughout my adult life, thanks to a college friend whose aunt and uncle studied at Penland, a nearby world-class craft education center, as glassblowers. Like many artists, they decided to stay after their workshops were over. In a clearing not far from here, I once attended a party where someone had a two-story-tall puppet, built just for fun. I've eaten pizza from an earthen oven crafted from the ground it stood on. Everywhere, there are gardens. Everywhere, there are carless trails between houses. It is, in my experience, a community that has done exceedingly well with the utopian concept.

Celo was founded in the 1930s and is run as a land trust governed by residents. The original community constitution states that the collective's purpose is "to provide an opportunity for its members to enjoy a life that includes personal expression, neighborly friendship and cooperation, and appreciative care of the natural environment."

Here, electric light isn't taken for granted. Like everything else that's incorporated into life at Celo, every addition and subtraction is carefully considered. The artists and farmers who caretake Celo's 1,100-acre property recognize the land as having its own needs and autonomy. Residents own their houses, but the land itself is never sold; it is assigned for lifetimes. Like Tal walking backward, Celo fosters a culture that doesn't do things just because they're mainstream or expected. They do things by consensus, always with respect for other living beings.

The houses tucked away in these woodlands are a mix of new and old. Not many of them have the log-cabin vibe people associate with Southern Appalachia—a region that is a diverse and ever-changing mix of rural and urban, new and old. Celo is a place of living history, a wink to Appalachia's artisan leanings. It was founded as part of the back-to-the-land movement that followed the Great Depression and mimicked some of the traditions familiar to my maternal grandparents. They were born and raised in families who'd never left the land for cities in the first place, surviving as subsistence farmers on what they could raise from the ground and salvage from industrial leavings.

Making my way through Celo to meet Tal earlier, I saw a group of children frolicking in a field with a milk calf. Here, roads have names like "Clay," inspired by potters who turn earth into dinnerware. A meal on these grounds is more likely to include lightly sautéed asparagus than my grandparents' vinegar-soaked beans. But if you took my family's old-time folk art and farming traditions full circle, you'd meet Celo's rootsy avant-garde traditions. I think this is why, to me, Celo has always felt like a version of home.

Somewhere in the distance, a rooster crows, heralding dusk rather than dawn. Tal leads us across a small hand-hewn bridge to the side of a road. This allows us to bypass the creek that cars must quickly drive through if they want to travel beyond. This is a road that's been built on the waterway's terms rather than in accordance with human desire. The footbridge is covered in mesh intended to give our feet traction. We walk across boards that have been nailed in crooked, creek water flossing between them.

When Tal started giving tours, years ago, inspired by blue ghost populations that appear at this spot in spring, he had a vague understanding that his eyes dilated to let in varying degrees of light depending on his environment, like a mechanical camera. But as he's spent more time outside, he's become more interested in the complexity of what's happening in his body.

People generally tend to think that their eyes have acclimated to the dark after a few minutes. But while it's true that the fastest gains of night adaptation are made within the first 30 minutes of light reduction, studies at the University of Illinois reveal that it can take several hours without flashlights or phone screens for eyes to reach full sensitivity, which, at its peak, grants humans night vision that can be one million times more powerful than what we utilize in daylight.

Tal adjusts his white baseball cap, chosen for low-light visibility. "Have you ever heard of rhodopsin?" he asks. "It helps us see in dark." Silence. Crickets. As it turns out, we've come hoping to see external, nocturnal wonders only to realize that, in some fundamental ways, we don't know how our own night vision works.

According to Tal, rhodopsin is a purple-pink-tinted protein that my body has been generating since light began fading this evening. It was first observed in the 1800s and named for the Greek roots of "rose" and "sight." At night, humans have access to rose-colored lenses through which to view the world, though we don't often realize it. Rhodopsin facilitates a biochemical reaction that allows humans to see in low light by helping to convert photons—that is, the energy coming at us—into electrical signals. And it's a process that takes time.

Years ago, before the development of navigational technology, airline pilots used to wear red glasses for hours before flying at night to take full advantage of their night-vision abilities, because red light does not deplete rhodopsin as much as other wavelengths, though all artificial light instigates at least a slow siphoning. Even though humans cannot see as well as some animals built for nocturnal living, many of us have more night vision than we realize.

Healthy eyes have, front and center, cone cells that assist us with making out color and detail. Our rod cells, located to the sides, are better at operating in dim light, and rhodopsin is a rod assistant. This is why, at night, we often catch things out of the corners of our eyes. "If rhodopsin is exposed to light, it gets bleached out and your body has to start making it again," Tal says.

I think of the pain of light hitting my eyes in the dark as the sting of bleach, an actual erasure that forces my body to start a physical process all over again. I think of all the times this has happened: the burn of a flashlight, the flick of a switch leading to pain, the way exposure to artificial light after spending time without it delivers the sensation of a lush velvet darkness suddenly stained.

Across a summer-buzzing meadow, we can make out the profile of a small house. "That's the last cottage we're going to see," Tal says. Before we enter full evening shade, which will speed the arrival of darkness, Tal pauses and pulls a roll of glow-in-the-dark stickers from his backpack. He hands them out and asks us to put them on to help the people behind us maintain their bearings. "These are not bright enough to alter your night vision," he says. "They'll just allow us to keep track of each other."

We press the stickers against shirts and backpacks as Tal delivers one last plea: "I need you to really absorb the fact that there can be no photos tonight. The light of cell phones is enough to undo everything." By that, he doesn't mean we'll harm the glowing mushrooms we're here to see; he's concerned that we'll damage our own night vision.

Tal is prone to include quotes in his emails and tour notes. "The world is full of magic things patiently waiting for our senses to grow sharper," his online bio reads. When I first saw the quote, pasted into our early correspondence, I'd thought it was a lovely bit of poetry. Now I recognize it as potential epigraph for a night-vision manual. Because sharpening senses isn't a matter of mental will; it's about giving them the space they need to develop on their own. In the modern world, this is harder than it seems. And difficulty continues to increase.

"Even smartwatches stand to be an issue out here," Tal says. This is the first season he's found their tiny screens to be a problem. Every new form of technology presents a threat that isn't always recognized in time to be thwarted. On night walks, smartwatches are not just founts of flowing information; they're tiny bottles spilling rhodopsin-altering bleach. I'm not wearing one, but several people in the group unbuckle the devices strapped to their wrists and slide them into their pockets. It's hard to process the reality of these everyday devices as capable of altering our body chemistry, directly and immediately.

Tal has the svelte frame of a trail runner. He knows this land well. But artificial light is something that cannot be outrun, and if a single indiscretion is made, it has the potential to take everyone down, sight singed. "I need you to be really careful with your night vision this evening," Tal says. "Protect it. Consider it a precious resource, because that's exactly what it is."

~~~~~~~~~~

On the forest-lined trail ahead, I can make out the shape of a mushroom. "Is that glowing?" someone asks, shocked that we've already had success. But it isn't bioluminescence; it's just a pale mushroom that's reflecting what's left of daylight. The confusion signals to Tal, who's still walking backward, that it's time to turn, forward facing, to greet the night. Mid-pivot, he plucks the mushroom like a flower. "You can say 'fung-ee' or 'fun-guy,' either works," he says.

"This is the fruiting body of a bolete. These are fun ones." Boletes are unique in that most mushrooms have gills on the undersides of their caps, but these have spongy surfaces with pore openings that disperse spores. "Every mushroom is pretty much a spore-dispersal mechanism," he says. "The rest of the organism lives underground."

Mushrooms are the fruiting bodies of mycelia, rootlike fungi under Earth's surface. There are forms of mycelia that never fruit, but a mushroom is always the visible manifestation of larger organisms at work. "You can think of the fruiting body of mycelium like an apple," he says, as if the larger, underground structure is a buried tree. Only from branches, we get the crisp tart of Granny Smiths. And from branching mycelia, we get the oft-edible fruiting bodies of fungi.

A 2023 study found that at least 59 percent of the planet's species depend on soil as habitat, and it's thought that almost every plant on Earth has a relationship with some form of mycorrhizal network. These networks connect individual organisms, transferring water, nitrogen, carbon, and minerals among them. Toby Kiers, an evolutionary biologist, has called the fungal networks a "global blind spot in conservation and climate agendas" in explaining why mapping these networks is of growing interest.

Generally, a single handful of soil can hold veinlike mycelia that would stretch for 60 miles. Studies have confirmed that trees, connected via fungal networks, work more like functional communities than competitors. Mycelia attached to tree roots can transfer nutrients, channeling resources from tall trees in sunlight to smaller trees that live in shady areas. Saplings that receive energy from those underground networks are more resilient to the stresses of climate change. And recent studies have shown that mycelia use electric signaling to communicate, with messages that move back and forth in conversation.

Foxfire's living light can come from mycelia, fruiting bodies, or both. There are even some species with bioluminescent spores, which send out reproductive materials in glittering clouds. New fungal species and behaviors are discovered all the time, all over the world. It's estimated that 98 percent of fungi on Earth remain undescribed. "It's still mostly a mystery how it's all connected and what everything is doing," Tal says.

He holds out the bolete's floppy cap. I press my fingers against its pores, pancake spongy, as he explains that fungi—bioluminescent or not—are heterotrophs that cannot produce their own food. Saprophytic fungi are responsible for breaking down organic material—plants, animals, and leaf litter—acting as a forest-wide digestive system that processes nutrients from dead organisms in ways that make them accessible to the living. "Even live trees have some dead wood in them," Tal says. "When you see mushrooms growing out of wood, they're often saprophytic."

He is standing next to a straight hardwood trunk. "Lots of animals, including humans, are more genetically related to mushrooms than to this tree," he says. "It's all relative. Or rather, we're all relatives." Humans have been found to share roughly 50 percent of their DNA with fungi species.

As we move on, Tal reels off lists of things people often mistake for foxfire in full light, including green fungi, which gives rotting wood the appearance of turquoise. The mention makes me realize that, just days ago, I came across green fungi and wondered if it might shine. But bioluminescent species don't usually provide hints to their unique attributes in sunlight.

Some people in the group have come to walk with Tal because they live in cities where both darkness and forests have become hard-to-find resources. But other regional residents like me have come out to follow him through darkness not because Celo has species that our communities do not but because Celo—oasis of land-based culture, as well as mature forests—practices engaging with these mountains in ways that we have, in the span of just a few generations, mostly forgotten.

At this point in history, a full third of human beings on this planet can no longer see the Milky Way from where they live. It is not a stretch to say that, at current pace, there will come a time when humanity will not be able to see stars at all. And though foxfire fungi might not be as sensitive to light pollution as fireflies, our bodies cannot perceive foxfire's light without darkness.

Already, in urban areas, there are children for whom stars are only phenomena mentioned in fairy tales. For them, it takes faith to believe that those celestial objects are real, visible elsewhere. Difficult as it might be to accept, and despite the fact we rarely talk about it, there is a good chance that we are among the last generations that will be able to see the night sky. Even less considered is the fact that this means we might be among the last to live in a world where natural darkness is deep enough to reveal the presence of foxfire and glowworms and fireflies. If light pollution does not abate, lots of species that already seem too magical to be real will, along with starshine, fall into legend.

Learning to Walk

"How's everybody feeling about walking in the dark? Are you getting the hang of it?" Tal asks. He's answered by a chorus of murmurs. We're making it, but we're not exactly sure of ourselves. The sky still has a sheen, but we can no longer see where we're putting our feet.

"Stay steady and keep moving," he says. "Move heel to toe." Solid ground is no longer something I can take for granted.

I'm paying attention to which part of my foot is touching earth first. Heel, toe. Heel, toe. The technique is preventing me from tripping over all sorts of things.

I'm in a heel-toe rhythm when I see a hint of light alongside the trail, inspiring an abrupt stop. It's the unmistakable outline of a fallen branch. Tal navigates a stand of trees to retrieve it. We trace his approach via the glowing green wand that's floating toward us. We've found our first foxfire species, bitter oyster.

"Bitter oyster is hard to see from a distance, but it can get pretty bright up close," he says. "These are fresh."

"How do you know?" I ask.

"They're soft and slightly sticky. You can touch them if you want," he says.

The wand floats toward me. I extend my hand to touch its end. The foxfire is rubbery, soft enough to yield. It leaves my skin coated like pine sap, though the residue filling the grooves of my fingerprints isn't as thick. "You can break off a bit if you want," he says. But I already have, accidentally. The bitter oyster is so sticky that pieces of its fruiting body have attached to my skin. My hand is encrusted with tiny jewels alight.

I touch the wand's glow, again and again. Each time, a tangy scent is emitted. I could place this bitter oyster on my tongue. It's not known to be poisonous. But I don't have to consume this to belong to it.

"It won't be long before these bitter oysters begin to age," Tal says. "They harden, they dry. They only glow at certain stages."

Once you've discovered a magic wand in a dark forest, it feels wrong to discard it. But Tal promises this is only a hint of fox-fire's abundance. So we leave the branch on the ground where we found it.

It's wondrous to know that the innards of the wand-stick are full of light threads, unseen. My mind moves out, like branching mycelium, thinking about how every fruiting body is just an outer manifestation of inner-spreading beauty, something greater than itself. .

In Oregon's Malheur National Forest, there is a honey mushroom known as the Humongous Fungus. It is arguably the largest living organism on Earth. It spreads primarily underground, with mass not visible from Earth's surface. It has been alive for more than 8,500 years, growing under the forest floor. Almost every article I've ever read about it has omitted what seems, to me, its most fantastical aspect: The Humongous Fungus glows, producing subterranean light for miles.

Every year around this time, it produces fruiting bodies that indicate its otherwise-shrouded presence, as does another fairly humongous fungus in Michigan. That one has been alive almost 3,000 years, though it wasn't known to science until the 1990s. These giant fungal networks are vastly older and larger than the trees they connect.

Bioluminescent mycelia—and, really, mycelial networks in general—only reveal themselves in bits; to underground grandeur, fruiting bodies are only hints. Right now, we're walking past dead tree stumps potentially full of living light. And there's no question that there are fruiting bodies all around us, on the verge of emerging as clues to worlds we've only just started to consider.

I ask Tal if he thinks there are giant bioluminescent mycelial networks in our home region. "It seems reasonable to think that we have whole acres around here that are glowing underground," Tal says.

For all the elementary school science book pictures I've seen of Earth's molten core, I remember no lessons about luminous tattoos just under the planet's skin. I don't recall ever being told about the networks of living light that are reaching and yearning and making community to support almost all the verdant plant life that surrounds me.

As we hike on, moving tenuously as tightrope walkers, I'm oddly comforted by the notion that foxfire might be knitting a net of light below me, growing and glowing as it has since a time before human religion, or language, or memory began.

~~~~~~~~~~~

Before long, we've reached a crossroads. "You've probably gotten used to the road by now," Tal says, "but the trail we're going to take from here is narrow. It's different, more challenging."

We have no guiding light, only the direction of Tal's voice. "We'll need to travel one by one from here," he says. We line up behind him, single file, our glow-in-the-dark stickers in alignment. Then we disappear into a stand of rhododendron.

The tunnel of rubbery leaves is several degrees darker than the already dark landscape. "I'm going to tell you when there's a dip in the trail, and we've reached the first," he relays to me, since I'm directly behind him. "This is a big one," he says. "There are also roots. Let the person behind you know."

I switch up my heel-toe strategy, shuffling. I slide my way across one tree root, then another, gently lowering myself. I'm glad I chose shoes with flexible soles. It's a strange sensation, to ascertain the shape of what's underfoot as important, my feet groping at the ground like a second set of hands.

"Root," I say to the teenager behind me. "Dip."

"Root," she forewarns the next person.

There's a minuscule dot of light pulsating on the side of the path. "Likely firefly larva," Tal says. He sees larvae on the forest floor all the time, sometimes even into winter. But there's no time like the cusp of fall for finding foxfire. "The 'fox' in foxfire," he says, "comes from the French term 'faux,' fake."

Tal slows his pace. "Do you always walk this route without lights, even after your tours are over?" I ask.

"Yeah," he says. "Walking around in the dark is how you find new things."

When Tal first came to Celo, he was a teacher at the on-site school, a place that doesn't allow cell phones. It's a rite of passage for students to walk home in complete darkness. In solidarity, Tal often joined them. It was on one of those walks—traversing the unknown alongside his students—that Tal first saw blue ghosts and realized they were something worth sharing with others. And it was on a walk home after leading one of his blue ghost firefly excursions that he first noticed the species he's taking us to see, an entire tree alight with mushrooms potentially undescribed to science.

It wasn't until he took photos and posted them online that he realized he'd found something others didn't know about. "People started commenting," he says, "and I realized the species was really unusual."

"Is it edible?" a guest asks.

"It's so small, you wouldn't even know you had it in your mouth," Tal says. "You'll see what I mean when we get there."

A rhododendron branch taps my shoulder to let me know that I've strayed from the path. I sense Tal stumbling in front of me. "Dip," he reports.

"Dip," I call.

"Dip, dip, dip," the voices behind me ding, as each person in line warns the person behind them. Without sight, we're mapping the trail through sound. It's an exercise in trust, with people whose faces we've never fully seen. And it is a lovely, unusual feeling, having human sounds, rather than artificial light, guiding us through nocturnal terrain.

I ask Tal if he's seen a change in people after these walks, peculiar as they are. "In the literal sense, I don't really see people at all," he says with a laugh. This is, to him, one of the most interesting things about night walks. After greeting strangers in low light, he typically never sees them again, though they almost always go on to spend hours walking around together. Still, those unseen have presence. He quickly learns the cadence of their voices, the character-revealing insights they share from the dark, which grants each attendee a shroud of privacy, even when enmeshed in a crowd.

The most memorable transformations he's observed have taken place among teenagers. "I've had kids terrified of the dark, I mean, shaking with fear. And once they see bioluminescence, they become entranced," he says. "They shed fear; it's like they altogether forget it's something they were carrying."

In front of me, I sense a tumble. "Root!" he shouts.

"Root," I say, turning my head to amplify the alert.

"Root, root, root," vocal cords ping.

In Victorian times, people marked trees to guide them home. They dug braille-like pocks on trunks to help them feel their way along. It's an intimacy of place that Tal seems to have without causing tree damage, so acquainted is he with this landscape. And it is his familiarity that's allowing us all to travel freely.

I suppose, in daylight, I have familiarity with my specific plot of land, too. Years ago, a friend told me he knew old-timers who could gauge the depth of the New River by where currents hit a certain rock. After nearly two decades of river living, it's something I can do, too. And there is more than one life event that I mark by the blooming of rhododendron, which drops blossoms in the river around the first of July. Those petals on water are, for me, more effective reminders than notes on a calendar. The intimacy I have been seeking with night is one that I've spent a lifetime casually cultivating with these mountains in daylight. What a gift it is to realize that I'm pining for more of what I already have, a deeper connection with the land on which I already stand.

But tonight, the familiar wilds of these mountains have been made, once again, strange. We're still a good journey away from Tal's fungal discovery, but we pause when we see a puddle of ethereal light beside the trail. It's surrounded by flecks of light that speckle the ground like paint splatter. Tal explains that it's mycelium in the process of consuming fallen leaves. "If I turned a flashlight on," he says, "this would just look like ordinary leaf litter."

Tal picks up a handful of light and gives it to me. The leaves are cool and damp, well on their way to decay. "Mycelium is really running right now." Running, like a stream in motion, like a living river.

I cannot see the teenager behind me, so I hold the glowing pile of mycelium up to announce it as an offering, inhaling wafts of the leaves' mild, medicinal scent. She is the youngest of the group, and she stayed in her father's car until the minute we left the parking area earlier, cell phone light pointed directly at her eyes. In the dark, I've heard her emit a few grunts indicating that she'd rather be somewhere else. But, in this moment, she comes across as wholly present.

She seems to be carrying a fear of missing out on cell phone time more than she's exhibiting a fear of the dark, but she sighs when she accepts the mycelium, as though she's traded something heavy for the puff of light. "How beautiful," she whispers.

We are, in essence, watching death become life. We are watching darkness being transformed into light. The teenager stares into the handful of alchemy before, finally, passing it on. "Mycelium," she announces, letting the person behind her know what's coming, announcing wonder as she had previously passed on warnings.

"Mycelium, mycelium, mycelium." Each handoff is announced, connecting us to each other. We are no longer faces. We are no longer hands. We are all part of the darkness. Because, in this forest, the glow of this leaf litter is all that visibly exists. It moves through the night, tracing the path of our voices, like a luminescent cloud pushed by invisible winds.

Tiny Lanterns

The farther we travel into the woods, the soggier the ground gets. It's clear from the darkness of the overstory that we've entered areas that don't see much sunlight. One minute, we're in complete darkness. Then a bioluminescent tree appears. Tal doesn't need to point it out. His main task was leading us here so that we could discover it ourselves.

This is something beyond handheld marvels; it's an entire tree trunk covered in a glow that no one knows what to call. The light looks, to me, slightly bluer than the green light of bitter oysters. We cannot see the shape of a single mushroom from here. They're too small. But we can discern the profile of the tree they're growing on.

Each mushroom appears as a chink in glowing armor that's covering an ash tree. This concerned Tal at first, because emerald ash borers—invasive insects native to Asia—have already decimated millions of ash trees in North America. Borers might even have played a role in what we're seeing, since the dead parts of this tree are what fungi are feasting on.

If the species was dependent on ash trees and they were lost to borers, this foxfire might be gone before the mushrooms creating the dreamscape could be given a name. Fortunately, Tal has found what he believes to be the same mushrooms on another tree species, which means they might not be choosy.

The light of the tree draws us in, and our single-file line turns into a mass of awkwardly bumping elbows as we try to figure out where to stand. My careful navigation of the path is forgotten. I am so focused on the lights that I end up with branches in my face. The tip of a limb slips through my lips, hooking me like a fish. I sputter and wave my hands erratically to escape.

Branches guide me to the ground where, on the tree's north side, I crawl on hands and knees, until I'm nearly close enough to feel the tree's moss-bearded face against my cheek. I cannot see individual dots of light until I am less than an inch away. There, I can make out capped mushrooms that shoot out of the tree to rise upward like umbrellas the size of pinheads.

Tal has sent samples of them to a mycologist who is attempting to help him sleuth out where the fungi fit. "Figuring out how to get DNA in a way that might work for identification has been hard," he says.

Tal's contact suspects that it's an undescribed species. But there's a chance it's a known species that wasn't previously known to glow. Or that it is known to glow elsewhere and is a surprise resident of the region. "If it is an undescribed species, who gets to name it?" someone asks.

Tal isn't sure. A male voice calls out a suggestion: "The Tal Mushroom!"

Immediately, Tal rejects the idea. "It bothers me when I'm learning about something and it's named after a human being," he says. "I don't think species or places should be named after people. All species have their own attributes. I like it when things are named after their characteristics. The blue ghost, for instance. That's a name that helps you recognize it when you see it."

"So what are the attributes of this?" I ask.

"It glows; it's tiny," he says.

"Maybe something like 'tiny lantern'?" I say. "As in: 'Look! Over there! I think that tree is covered in tiny lanterns, first described in the Celo community of western North Carolina!'"

"There you go!" Tal says.

As we walk on in darkness, it seems the conversation has been forgotten. But the forest is so quiet, outside of the scraping of our soles against rocks and roots, that I can make out the sound of Tal whispering to himself in Latin. "*Minima laternis*, tiny lanterns," he mumbles.

~~~~~~~~

We step from soft soil onto gravel, where my feet turn broken granite. The overstory opens, and the encompassing darkness we've been treading is replaced by cloud-filtered gloom. "Is this light just because we're not in the shade anymore?" someone asks. Tal confirms and says, "Some of this might be light pollution. And the moon is higher now."

I have, many times in my life, gone out of my way to watch a sunset, but I've never really thought about the timing of moonrises. Before heading to Celo, I glanced at my moon calendar to see what phase it might be in this evening, but I didn't take its movement into account. But Tal did, and he planned this walk so that we'd emerge from the woods just as the moon was crowning mountains.

"You know," Tal says, still ruminating over our conversation about tiny lanterns, "I think it's good to call species 'undescribed' rather than 'undiscovered.' I'm not the first person to find those, I'm sure. There were people who knew about that species at some point in history," he says, acknowledging that we're walking ancestral homelands of the Cherokee. "The thing is, most of us don't see the world around us, not even in daylight anymore, really. Plant blindness. You've heard of it?"

The term was coined in the 1990s as a way of talking about how people—particularly those from traditions that do not incorporate plants into spiritual and cultural practices—have a bias to overlook and underappreciate specific plant species. Studies have found that plant blindness, which is sometimes referred to as plant awareness disparity, alters how greatly people care about conservation efforts, with less knowledge leading to less concern.

Standing in the night shade of Celo's community forest, it's hard not to wonder if light pollution is being allowed to increase at exponential rates because, as a whole—particularly in parts of the world where artificial light is most overused—humans are experiencing a similar, relational disconnect with darkness. We depend on natural nights just as we depend on plants. We have become not only species unaware; due to the endless faux days we've created, we've now reached a point where we hardly notice that night itself has its own characteristics and functions.

Plant awareness disparity isn't as great in some cultures as it is in the United States' mainstream. Studies have shown that in India and Sweden and diverse indigenous communities around the world, the spiritual, emotional, and practical relationships people have with plants encourage connection. It might not be a coincidence, then, that I was inspired to seek out the nocturnal wonders of foxfire because *Foxfire* has, for all my life, been something associated with subcultural Appalachian folkways— including things like planting gardens in accordance to moon cycles—familiar because they've been practiced by my ancestors in these mountains for more than seven generations.

Humans have been found to be more adept at identifying animals than plants. Plant awareness disparity is thought to be a chemical and visual bias of the human brain—which, overwhelmed, tends to lump what it cannot easily process—but it's something that can be overcome with species proximity and cultural conditioning, as evidenced in communities around the globe. Seemingly in their own wisdom, foxfire species have presented themselves not as part of the relentless din of day, but rather as gems laid out against velvet night to be inspected as precious.

Our time with Tal is officially over, but no one walks back to their cars. Instinctively, we form a circle. Here, we have seen forms of light that relatively few humans ever have. Not one of us is ready to leave, not even the teenager, whose cell phone remains tucked away of her own volition.

"Being out here is a privilege," Tal says, acknowledging the rarity of having unfettered access to a place like this, particularly one that, unlike many public lands in the area, is relatively flat, lowering our chances of walking off a cliff.

Even though we've been wowed by fungi, it's the depth of Celo's darkness that seems a luxury. It is a reminder that, once upon a time, electricity was only for the rich and urbane. Asheville's Biltmore Estate, built by the Vanderbilt family, remains one of the largest privately owned houses in the country, and it was the first house in this region to have electricity.

For my own rural farming family, electricity is a recent reality. In the 1940s, when my grandparents were young, only a fourth of North Carolina's farms were wired. The electricity that runs through my house—outside of what's generated by an array of solar panels—is still delivered by a regional cooperative. Historically, the only way my home county could tap into the grid was by acting communally, since, for a long time, power companies refused service to rural, mountainous areas. But even the most remote regions on this planet are rapidly changing.

A 2022 study indicates that artificial light, once mostly an urban issue, is now also associated with nonurban populations, because light has started bleeding, often unmitigated. Rural areas that endure other forms of environmental pollution—including the degradation of air and water—are also, increasingly, being bleached by light.

The socioeconomic indicators of darkness even extend to the shelter of midday shade in cities. Inequalities in urban neighborhoods track a lack of trees, which, when isolated in asphalt seas, require human resources and funding to stay healthy. A 2021 study published in *Landscape and Urban Planning* found that a lack of trees—and thereby, shade—directly correlated with race- and education-based inequalities. Studies about artificial light pollution have similar findings, in reverse, with socioeconomic inequalities correlating with greater light pollution at night. In a warming world, vulnerable communities are being scorched, day and night. Whereas once, artificial light was rare, it's now natural darkness that has become scarce.

Tal's sister, a teacher visiting from Boston, says, "I wish my students could go on a walk like this."

"I wish everyone could go on a walk like this," I say. The group hums in agreement.

Tonight, as members of the group travel home, we will reestablish our dependence on artificial light as we start our cars. Before this, we might not have considered how headlights stand to negatively alter our vision. We might have erroneously assumed the brighter the lights, the better our sight. Recently, I learned that the pirates that populate childhood storybooks wore patches over their eyes not to hide some sword-fight disfiguration, as I'd long thought, but because they likely wanted to keep one eye attuned to the darkness belowdecks even at midday. They, like Tal, knew that fully developed night vision was something worth safeguarding.

The LED headlights that have gained favor in recent years can be up to three times as potent as traditional halogen headlights. In the United Kingdom, a recent survey revealed that two-thirds of drivers say they are regularly dazed by modern headlights, with 67 percent saying it takes up to five seconds for them to regain functional sight.

The United Nations has founded a World Forum for

Harmonization of Vehicle Regulations, but the United States has been relatively slow to embrace adaptive lighting. In 2012, the American Medical Association adopted a statement calling light at night a health hazard, citing concern about "disability glare," a reduction in visibility caused by headlights. Excessive light was deemed a road safety risk.

Those of us driving up and down mountains will soon lose much of our hard-earned night vision while chasing LED lights, but Tal won't. He'll be walking home through the darkness we've just emerged from, accompanied by rivers of light and spangled trees and other magnificent things, discoveries of which we cannot yet dream.

~~~~~~~~~~

Even though I live in a rural area—in the actual boonies of Boone—my neighborhood no longer knows Celo-level darkness. In recent months, a new storage unit has been erected on my side of town, its lighting scheme overlooked by ordinances. Meanwhile, someone on a hill across the river has installed a new LED porch sconce, which blasts like a maximum-security prison floodlight from their front door.

I bemoaned these changes, but I did not fully realize how greatly they were degrading the world around me until I saw them in contrast to Celo. Still, the patches of mature hardwood forests around me almost certainly hold foxfire. And I'm freshly inspired to scout shadows as caches of wonder.

Soon after Archer hears about my foxfire encounter, he decides to join me for a sunset walk so that our sight might develop as sunlight fades. We sit on crumbling logs, allowing our eyes the space they need to greet darkness.

After a bit of time has passed, Archer starts asking, "Is it dark enough yet?"

"Let's wait a little longer before we start walking back," I say. Already we've identified red chanterelles, valued as a culinary delight, as well as destroying angels, among the most lethal-if-consumed mushrooms. In these woods, sustenance and poison grow side by side.

An hour passes. Archer grows impatient. It's dark, but not dark enough to see bioluminescence. This wooded trail is a little farther out than the roadside where we witnessed blue ghosts. He wants to walk back to the house, but he seems reluctant to leave me alone in the forest.

When we start to explore on foot, I suggest that he should move heel to toe. Out of the corner of my eye, I see a faint flash of light. Then, again. But Archer, focused on the path ahead, has missed it.

The rods of human eyes, responsible for a great deal of our night vision, generally only need a few photons—maybe even just a single photon—to be activated. Their location on the outer edges of our retinas strategically helps us catch things coming at us from all angles. It's something we, as a species, have depended on for nocturnal protection and hunting practices. But in the modern world, these side-eye powers feel strange. Maybe because we're not used to valuing what's on the periphery.

It's probably why I thought I might have been imagining things when I first saw that neighborhood bobcat. It's probably why I've heard so many people question their own senses in darkness: *Am I imagining that?* We generally know our cones, but we're less acquainted with our rods.

What I'm seeing out here isn't fungi. It's too bright. I suspect it's firefly larvae. Still, even on hands and knees, I'm having trouble locating the source of light. I try different points of focus, utilizing the corners of my eyes. Still, nothing.

When the forest releases us into a neighbor's garden, Archer makes a run for home, but I linger. On the side of our gravel drive, again, I think I see something. I lower myself until dandelion leaves seem as unfamiliar and exotic as palm trees.

My fingertips press against stone. Then, something else. A firefly-to-be, larva that looks not unlike a glowing roly-poly. I place the larva in my palm and run home to announce my find. Archer comes when I call. He can see the larva glowing. But, already, he's lost the high-quality night vision gained on our walk. "Maybe I'll try again tomorrow," he says, shrugging.

He is eager to return to the movie he's already started watching, but I can tell he's disappointed that he missed out on experiencing this living light at full throttle.

~~~~~~~~

When I notice a stick covered in milky discs on a midday walk, I almost don't pick it up. But its bumps have the outline of bitter oysters, so I collect it, just in case. I know how to identify bitter oysters in the dark, when they are aglow, but I have no idea what they look like during the day.

I drop the branch off on my porch and forget about it until it's nearly dusk, when I realize that, if I'm going to be able to confirm that I've found bitter oysters, I need to let my eyes adjust. I sit on a porch swing, watching birds weave through distant trees. I could take this branch to a darkened space, but—even if I were to find a windowless room to reveal this branch's secrets—I cannot intellectually will my sense of sight sharper. I can only give my body the time and space needed to summon my abilities, seek environments that will no longer stunt them.

I open the front door and shout for Archer to come out. When he emerges, he is, again, arriving from a world of blue light. Only this time, he decides he'd like to invest in reclaiming his night vision.

"Give it a minute," I suggest. What I mean is: *Give yourself a minute. Let your vision ripen.*

He romps through the yard with our dog, Wilder. After a bit of time has passed, Archer's night vision is far from full capacity, but he's gained enough to see. He cups his hands around the branch so that he can have a darker space for viewing. Then, he shouts, "Foxfire! Radioactive green!" Bitter oysters' glow is dim, but it does look radioactive, no black light needed.

Immediately, Archer suggests that we should take the foxfire to my mother so that she might witness what we've found. It's a way to align the finds of dark terrain with our not-always-surefooted elder's level of mobility. And, thankfully, this branch is wholly portable. So off we go.

When we arrive at my parents' house, my mother ushers us into a bedroom. It's darker than her higher-altitude yard, where the skyglow of Boone is bright and there are no trees to act as shade-shelter. With window blinds pulled tight, we feel our way to each other. She has been in a dim-lit space, so it doesn't take long for her to catch her first glimpse of foxfire.

"That's amazing!" she says as Archer waves the branch around as though casting a spell. It is a very real evocation of awe. My mother, normally demure, is so delighted that she bursts into laughter.

She has been in possession of the *Foxfire* books for almost half a century. But, for all the *Foxfire*-connected heritage my mother propagates, she has never seen foxfire, the actual fungi, for herself.

When she reaches out to touch the sticky-fresh fungi, crumbs fall on a log-cabin-pattern quilt that my great-grandmother constructed from the discarded clothing of the people she loved, ancestors I never met. "A magical mystical gift!" my mother says of the branch. The foxfire does feel powerful. Even more powerful than it did to me in Celo, given that it is held by my mother.

For her, the *Foxfire* books were one of the first indications that growing up on a farm—where my grandfather raised poultry, whittled duck decoys, and upholstered furniture with factory scraps—provided a wealth of folk knowledge that was valuable beyond her homestead's fencing. What those books gave her was a sense that she should better honor traditional knowledge that, out of embarrassment, she might have otherwise let fall away. She's told me that, before she read those books, she'd always tried to get her peers to see her as modern, contemporary, anything other than a backward hillbilly.

I've always known Appalachia as more valuable than the sum of its parts because my grandparents were masters of conjuring its magic. Long before I was introduced to the importance of oak trees as moth habitat, I knew the feel of white-oak splits softened in water—my fingers guided into basket weaves by my grandmother's hands. Since childhood, matriarchs have been showing me how to use needle and thread, but I've never had much talent for *Foxfire*-style crafts. My sad attempts have, at times, made me feel a little lesser than. I've always been more attuned to natural patterns than sewing patterns, and it has taken me a lifetime to understand that this is not a failing.

The biodiversity of the night world my grandmother knew no longer exists in full, but I have been trying to collect bits of the darkness that she wore around her shoulders before it wears too thin for stitching. Because bearing witness to the natural world is a folkway that must also be practiced, lest it disappear completely in an era of pan-flashes and abstraction.

For Archer, the word "foxfire" might not immediately evoke a book series at all. The term will more likely, for him, always be the glowing mushrooms that are floating in this loving womb of a room. The ghostly light we're fixated on is both life and death, past and present. And the memory of this evening will tie him to these mountains, sure as a quilting stitch—turning darkness into a familiar comfort that he can wrap around his shoulders wherever he's able to find it.

In time, my mother's declarative *oohs* and *aahs* about the foxfire begin to soften. Archer seems to realize that it would be easy for her to mistake initial novelty as all that exists. But he now understands that time can deepen the value of darkness as an experience. "The longer you stay in the dark, the better you'll be able to see," he tells her. Slowly, our nearly forgotten powers rise to meet our newfound patience as—hand to hand, generation to generation—we pass an actual torch of natural heritage.

Fall

Moon Gardens Blooming

~~~~~~

The Language of Flowers

The mountain growing season is short, and autumn is already tossing yellow-and-red confetti. It slaps against my windshield as I drive backroads to reach my friend Amy's homestead, circa 1892. Curve after curve, I find locust trees that are a few shades lighter than they were last week. Buckeyes also seem well on their way to change. It is, already, hard to tell the difference between orange leaves falling and monarch butterfly wings rising. The signs of summer and fall, all intertwining.

Amy, a hobbyist flower farmer, has invited me over for a one-on-one garden party during this early fall week. Tonight, we're determined to stake out night-blooming flowers so that we might see them open in real time. To some people, this might have the thrill of watching grass grow, but we've been looking forward to it for weeks.

Our admittedly quirky plan can be traced back to a conversation that happened months ago, when I told Amy that the University of Cambridge had livestreamed the nocturnal blooming of a moonflower. There are many species called "moonflower." The one broadcast from the Cambridge greenhouse was an imported cactus.

People gathered around screens to watch that Amazonian flower bloom in a European greenhouse, in a petal explosion that lasted till daybreak. It was heralded as a special occurrence. And once I heard about it, I could not stop wondering what it might be like to watch a flower rise to greet the moon rather than the sun—only in person.

Amy—the kind of woman who spins yarn from the wool of her own sheep and grinds wheat berries to bake her own bread—suggested that, while she found the public interest in that cactus flower heartening, she was wary of people gathering around blue-lit screens to witness an organic happening. "There's an entire night-blooming world out there," she told me, in commentary that inspired her to have a personal epiphany. Because even though she has dedicated much of her life to tending flowers, she's never pointedly set out to observe their nocturnal behaviors.

Just as butterflies have whole sun-blooming landscapes, moths have whole moon-blooming ones. I'd been introduced to the concept at Mothapalooza, but news reports of the Cambridge bloom made me start thinking about it anew. As Amy suggested, near each computer screen that had transmitted the image of that flower, at points around the globe, there were likely a variety of species revealing their own nocturnal beauty, unappreciated.

Moon gardens—with plants curated to be enjoyed after sunset—are designed with night bloomers and silver-and-white foliage meant to catch moonlight and ooze perfume. Many chosen species don't bloom just once a year, but every evening, for as long as the flowering season lasts. They apparently enjoyed a surge of interest during the pandemic, when people were stuck at home, searching for nightly entertainment.

Historically, moon gardens gained traction in Victorian times, when households were beginning to transition from candles and lanterns to electricity. In that era, darkness was not the other; it was endless light that was foreign. And if those long-ago moon gardens were a way of exploring people's growing relationship with light, how might modern ones stand to deepen our relationship with disappearing darkness? It's the kind of thing Amy and I couldn't stop wondering about once we got started.

One of the best-known moon gardens in the United States is, ironically, located outside of Thomas Edison's laboratory in Florida, where the scent of his cigars is said to have often mingled with the perfume of gardenias. Edison drew inspiration from that garden, cultivated in partnership with darkness. Yet we have become so acclimated to the electric world that he helped build, we have forgotten the value of the one it replaced.

This evening's flower-watching might seem eccentric in the electrified era, but it's a practice that's occurred throughout history, across cultures. In general, though, if you're not a gardener or regular night walker with an interest in botany, finding nocturnal bloomers can be difficult.

Almost as soon as I heard the term "moon garden" I started putting out the word to see if anyone I knew had one, becoming particularly romanced by the idea of tropical moonflowers, large ones like *Cereus* and *Ipomoea alba*, which resemble the morning glories that once grew feral along the edges of my grandfather's tar paper barn. Then, one day, I unexpectedly heard from Amy, who, after our initial conversation, realized she'd sown some night bloomers that were maturing.

She'd been drawn to the ornamental tobacco genus *Nicotiana* not because of its nocturnal status but because she had identified it as a collection of plants that would have traditionally been grown on her Appalachian homestead. It was only after contemplating that live-broadcast bloom that she'd started thinking of watching flowers minute by minute as a potential recreational activity. Now she's romanced to think that it's something the original inhabitants of her home, known as Honeysuckle House, might have participated in.

Amy lives, in part, like a historic reenactor, pursuing house projects and activities that align with her home's original timeline. She would hesitate to describe herself as such, but it isn't just anyone who mows their yard with an antique scythe, storing it in the corner of a room known as the parlor. She has, since I've told her about this revived Victorian trend, come to realize that her historically accurate plantings have led her to create a haven of almost-lost nocturnal pleasures without realizing it— simply by tracing the inspiration of generations who lived prior to electrification.

Still, Amy wasn't convinced she had anything special to offer after dark—until she spotted her first hawk moth. I'd told her that tobacco species are, as night bloomers, often frequented by hummingbird-like creatures that coevolved with *Nicotiana* and other night bloomers like primrose, the shape of the moth's proboscis tongue almost an exact match for their tubular flowers. As soon as she learned about hawk moths, she started going into her garden at night to look for them, and immediately, one showed up.

The encounter, nothing more than a fleeting glimpse, was so moving that she nearly cried. Reflecting on the emotionality of her response, she said, "I didn't even know hawk moths existed until you told me. Then, when I went looking, I found them, right there in my garden. I still can't believe it."

I get it. I have, for several seasons, been living it. Is there anything more moving than awaking to wonders that you have been wandering among all your life unaware? Is there anything more hopeful than realizing that you've always been surrounded by sublime scenes, even when you were living through days and weeks and months full of despair? Once you've brushed against night's magic, it's hard not to yearn for more of the shimmering life that seems to reside in all the darkest places.

As soon as the moth convinced Amy that she had nocturnal bounty to share, she called me over for this mini moonflower party. We've heard that, if you catch *Nicotiana* at just the right time, it's like watching a time-lapse video—only with three-dimensional senses. Given the small seasonal window, we've made a pact to stake out these flowers for as long as it takes to bear witness.

~~~~~~~~

When I arrive at Amy's gingerbread-trim Queen Anne Victorian, she isn't basking in the glory of her gardens, she's gazing into the walnut trees that stand watch over her night bloomers. "Vultures," she says, without diverting her gaze, which is sharpened by eyeglasses round and wide as Mason jar rims.

I see one, then two, then a whole committee of vultures staring down from a crooked branch. "They've been here for a while. We're going to need to go check on the sheep." This is not the flower-full experience I was expecting. But, inspired by my own adventures with laying hens, I've long stated: If you don't have enough drama in your life, get some livestock.

I slide on the borrowed muck boots she offers, and we orbit a constellation of outbuildings to reach the pasture. Past a white-washed chicken coop. Past the clapboard apple house where fruit was once stored through winter. Past the spring house, where drinking water was scooped from a trough prior to indoor plumbing. All the while, at least five vultures are watching us.

I'm trying not to take it as a terrible omen that they are perched directly above Amy's kitchen garden, where the night-blooming tobacco is growing. We were set to look for signs of life. Now we're faced with the very real possibility of finding a carcass. It's an ill-matched introduction to flower-watching. At least, that's what we're thinking as we climb into her paddock, where scavengers are circling.

She lifts a creaking cattle gate, and we climb the pasture's highest elevations. During this golden hour, the rolling hills are resplendent as a crown, with a mountain valley blushing pink underneath. It is the view held all day every day by her flock of Scottish Shetland sheep, bred to thrive in harsh conditions that mirror those of our home. When we crest the top hill, woolen heads raise to register our presence before going back to grazing.

"Looks like they're all okay," Amy says before turning her eyes skyward, where birds are still circling. They're so large and the sun is so bright that their bodies cast shadows like art mobiles drawing circles on the hilltop, over and around us. I turn my eyes to the ground, and I watch their shadows slip across pasture grass. This is as close as I've ever been to a churning vulture kettle.

There must be a reason the scavengers are here. But Amy's worried tone has mellowed. Now that she's counted her sheep among the living, she's convinced that it isn't death the vultures are seeking. "It's cattle birthing season," she says, nodding toward a line of barbwire that marks her neighbor's pasture, full of first-calf heifers. "I'm thinking that the vultures must be here for cattle afterbirth."

Death isn't always bloody, but mammalian birth is without fail. The nutrients of those mothers' dark-womb placentas are, by the calves that they supported, no longer needed. But they can still nourish the community of life at large. The vultures are likely here to turn the remains of those domestic births into energy that might support their own wild offspring.

Vultures are famed symbols of darkness and death. But today, their presence is heralding new life. It's a stunning realization to have, standing on this mountain bald, with vultures closing in. We came fearing a funeral, and we found a birthday celebration instead, on this, the very evening that we've decided to watch flowers—thought of as only blooming in alignment with sunlight—come to life in darkness.

～～～～～～

On our way down the mountain, Amy's sheep bleat a serenade that carries us through beds of cotton-ball hydrangea and swaths of color that she will soon sell via a local farm cooperative. When we reach the kitchen garden, we pull chairs across flagstone so that we might sit face-to-face with tobacco plants. Above, the vultures keep watch. They are still a bit eerie, but—given our strange realization about them as life-affirming creatures—I'm starting to appreciate their company.

Amy's garden includes a variety of *Nicotiana*. Some bloom in daylight; others at night. The day bloomers are already in full form. They include woodland tobacco, *Nicotiana sylvestris*, which is bursting forth with white flowers, sprays of living star showers. Their blooms are hanging so low it looks as though they're streaking white light across green stalks.

She has laid planks of weathered wood down to help her reach the far end of her beds, where various species are growing along fieldstone bulkheads. Of particular interest is jasmine tobacco, *Nicotiana alata*, with tight green stalks. These are the night bloomers we've pledged to watch.

As we wait out sunset, Amy decides to introduce me to her population of tobacco hornworms. She puts her hand under a leaf covered in nibble marks to reveal a green marshmallow. "When I saw the moth, I knew that caterpillars were around, and, sure enough, I immediately found this guy."

The creature's soft body has a sharp red tail that points up like a horn. I lift a leaf to study the caterpillar and gain sticky residue on my hands. If tobacco is harvested without gloves, it can create sickness from absorbed nicotine, a gluey mess on skin.

Amy directs me to smell the tops of flowers, then the bottoms. "It's weird, how different it is," she says. The petals have a sweet, delectable scent. But the stem smells distinctly of cigarettes. After a few extra sniffs, it takes on the heavy musk of leaf-rolled cigars, and I cannot help but muse that all those historic reports about Edison in his garden might have referenced the smell of some tobacco growing, not just the dead, dry leaves he was smoking.

It is, for me, a shock to view tobacco as a living being rather than a packaged product. North Carolina is so closely associated with growing tobacco that once, as a child in the 1980s, I was taken to a cigarette factory on a field trip, where whole packs of Camel Lights were given out as souvenirs. I was, maybe, twelve. Though it is nearly a bygone crop now, I have memories of burley tobacco growing around Watauga County, where it was long harvested and gathered in field bundles that looked, from a distance, like tents. Part of my current house is a reclaimed tobacco barn, salvaged for lumber. Some of our rafters are primitive tobacco drying rods, straight branches with bark still on them.

When Amy's tobacco first bloomed, she picked some for a bouquet, intrigued by their floral scent. But by the time she got them inside, the perfume was gone and all that was left was the musky smell that she'd always associated with tobacco. To her, as a child, even before the dangers of secondhand smoke were widely known, the smell of cigarettes always seemed offensive, dirty, gross. But growing tobacco has forced her to relearn the odor, create new associations. A scent she once connected with secondhand smoke has now mostly been replaced with more pleasant garden associations.

I lean in to smell the strangely dual-scented jasmine tobacco blooms, and when I get too close, flyaway pieces of my hair stick to the plant, as though it's attempting to pull me in. Briefly, it succeeds. But I pinch a flower stem and pull away, freeing myself to go sit in my chair like a vase.

Taking cues from the insects munching around us, Amy—a

host prone to old-fashioned hospitality—slips inside and returns with slices of homemade bread and cups of hot tea, sweetened with local honey. Her reappearance inspires the last of the vultures to fly off, leaving us to dine with warm-faced marigolds on a carpet of forget-me-nots.

As it grows darker, Amy notices that a light has been left on inside the house. She makes a *tsk-tsk* sound before going to turn it off. "Since I saw my hawk moth, I've been making sure to keep things dark," she says. Now that we're sitting here, awaiting dusk, she finds it interesting that she'd never really considered how artificial light might be altering the plants she spends so much time fretting over. Darkness has, after all, long taken over for her in the evening, safeguarding even diurnal bloomers so that she didn't have to water them periodically.

Amy is always thinking about sunlight, where she should plant things so they might maximize growth. She even owns a grow light that mimics the sun. But she's never really considered that pulling her living room shades down or turning off her lights would make any difference for her plants. But now it strikes her as headshaking-hard-to-believe that, though she's given an occasional thought to shade, she's never thought much about how natural night is cyclically required for flowers' photosensitive cycles.

Research on the overall effects of light pollution on plants is still sparse, but artificial light has been found to change entire grassland communities, with plants that respond more positively to electric lights pushing out other species. And some plants, particularly trees, leave their pores open for unnaturally long periods under the influence of artificial light, which makes them more sensitive to air pollution and drought. Masking natural night with artificial light can alter a tree's immune system, not unlike how lost sleep lowers human defenses. Some plants stressed by artificial light have been found to over-photosynthesize in attempts to match its unnatural energy, which creates more stress. Under artificial light, plants cannot properly rest. They do not have the time they need to process, like animals denied dreaming.

As we wait, perched on the edge of twilight, Amy tussles with her sheep dog, who is restless without herding tasks. I cannot sit still, either. I tap my feet, watch for vultures that might rejoin our party. But, in time, a strange thing happens—as darkness rises, our attention tightens. Our feelings that we should be taking care of something else disappear.

We focus on a single blossom. "Can you tell a difference? Are those petals moving? They're a little less green, right?" The petal undersides are lime green, but Amy assures me that they're going to reveal new things.

"When they're open they're clearly white, not green at all," she tells me. I cannot understand what she means. I have only seen what exists, not what is yet to be. Amy knows what they look like in full form, but she has never watched their process of becoming. She has only seen these flowers closed in daylight and open-hand-waving in the evening.

Slowly, then suddenly, one bloom takes the lead in opening. Each petal is relaxing, not stem to sky, but center-petal out. "They're curled like tongues!" Amy exclaims. "I never realized each petal had to unfurl on its own like that." Before, the flower was so sealed it was almost a bud. Now, it's loose enough for me to see that, though the bottoms of the petals are green, the tops are stark white.

These flowers are basically turning themselves inside out.

And they're doing it together, like synchronized dancers. One by one, they match the stance of the precocious flowers around them. Each motion fills the air with fragrance. Gone is the smell of cigars in a wood-paneled room. Rising is the smell of luscious floral perfume—the likes of which I've never experienced. This blooming tobacco is clean as evergreen with a spoonful of sugar on its breath.

We fumble and fail to find words to explain the way our bodies are reading the signals these flowers are emitting. I close my eyes, inhale deeply. Just as foxfire showed me that I have better night vision than I imagined, I can feel these night bloomers introducing me to my sense of smell as one that I have, thus far, mostly ignored as important. Every waft feels like gratitude for the sense of smell I've been taking for granted—maybe, as a species, most humans have.

A University of Manchester study suggests that we are losing our sense of smell due to genetic changes. Sight, the sense it seems we tend to value most—the sense we associate as being supported by artificial light—has apparently taken the lead in evolutionary progression. No one is entirely sure what's happening, but various studies about loss of smell, anosmia, have shown that air pollution—especially particulates from burning fossil fuels—might be partially to blame. Researchers in Italy, Brazil, Sweden, and elsewhere have identified a nearly twofold risk of developing anosmia when people live in areas with sustained air pollution. Anosmia has also, for years now, been recognized as a distinct COVID symptom.

Still, our collective sense of smell remains relatively acute in the scheme of things—more acute, some scientists say, than we often give it credit for. We can smell just as well as some dogs we admire as supreme sniffers; it's just that we're gifted at smelling different things. Even so, it's hard to get over: Our species might never again be this sensitive to the nuances of flower language on our evolutionary journey.

The scent of these tobacco flowers is so strong that it chases me around the garden. I ask Amy if she is experiencing a similar phenomenon, and she confirms that she, too, can feel the scent growing so thick that it seems it might, any minute, become visible, hanging in the air around us. Still, we're not sure if the flowers are producing more concentrated perfume, or if we're just becoming more sensitive to it. Maybe it's both, a sensory meeting of species.

Many night bloomers have enhanced, nonvisual characteristics to attract pollinators like moths and, in some cases, bats. They include fragrance, and the human olfactory system is connected to circadian rhythms that fluctuate in relative alignment with them. According to a 2017 study conducted by Brown University, human olfactory ability is strongest in the evening around 9:00 p.m. It is at the exact time nocturnal-blooming flowers begin to reveal themselves—with their scent levels deepening in darkness—that our senses peak to meet them.

In watching the behavior of these flowers, I feel like I am being instructed to remember that my sense of smell is not some add-on, something lower in hierarchy to sight; it is a fundamental part of how I engage with the world around me. It's a glory that screens cannot begin to transmit, so lovely that I can hardly stand it.

Amy directs me to touch the silvery cool of a lamb's ear plant she's growing near tobacco. It is commonly planted in moon gardens because of the way it catches moonlight in velveteen foliage.

She planted it because, as a shepherdess, she adores the leaves' texture, soft as the actual ears of a newborn lamb. Also, because Honeysuckle House's early inhabitants would likely have spent time enjoying this species in evening, after they'd tended to their own flocks.

When Amy notices a seed pod that's fallen from tobacco, she picks it up and motions for me to open my hand. Under the pressure of her finger, the pod breaks open and seeds fall freely. I press an index finger to where they've pooled in the center of my palm and roll the seeds of dark matter—hard and round as musket balls—with the tip of my index finger.

"Some seeds need total darkness to germinate; others need a little light," Amy tells me. "These will do best right at the top of the soil. My hope is that these plants will self-seed and that, left alone, they'll be carried by wind." But, given that I already have a handful of dark orbs, Amy suggests that I might spread these myself, as an offering. "Just toss them in there," she says. I roll the seeds across and out of my palm, grateful for the opportunity to participate.

Heirloom

It's soon cold enough for Amy to go inside to dig a jacket out of winter storage. In her absence, I can make out the rustling of sheep in the hillside pasture above. I am alone. Only, I am not. My eyes catch on a shadow. A hawk moth passes, quick as a paintbrush stroke.

The moth is gone by the time Amy returns. She's disappointed to have missed it, and I'm sad to have only seen it as a smudge. But we're both glad the creature has inspired us to lift our collective gaze, so that we might study the moon that has risen while we were focused on flowers. The silvery circle is precisely nestled between an exterior wall of Honeysuckle House and the ancient mountain that anchors it. The match of earth and sky is so perfect, it looks like a homespun Stonehenge. Amy settles into her chair, down jacket billowing. She's lived here nearly twenty years, but she's never seen the moon at this exact time, in this exact season, the way all these moving parts of nature sync. It's delicious.

In her day job, Amy is a psychology professor who teaches things like evolutionary and conservation psychology. Recently, she's been reading about a concept known as generational amnesia, and our moon garden experience is reminding her of it. "Basically," she says, "we get used to the way things are, and we forget the way they were. The baseline for what we think of as normal keeps changing."

Of her homestead, Amy says, "I love this place, but it's easy to forget that my pasture used to be forest. In the future, if the farmland around here is taken for development, people will forget that this used to be agricultural. Each time, we sense a loss. But over a long stretch, people lose perspective and can't remember what those losses mean."

As each generation inherits the world, we become increasingly forgetful of how things were before, and we forget how they might be. For better and worse, ever and always. We forget that the things we take for granted were, at some point in human history, new inventions. We accept that our sense of smell is what it is. We forget that landscapes we admire as wild or bucolic or supremely cultivated were, in the past, different.

Psychologist Peter Kahn, who coined the term "generational amnesia," suggests that this memory loss has generally resulted in an ever-downward shift in terms of what we accept as ecological health. One example often used is how elders can remember a time when driving at night meant finding the remains of hundreds of bugs on windshields. Now, insect populations have dwindled so much that, when night driving, you might not even find one. Yet, due to slow acclimation, we don't think much of it. And the next generation doesn't think of it at all.

When he conducted a study about how kids in urban Texas perceived the quality of their environment, he found that, though they understood pollution as a concept, they didn't realize that they lived in an area with serious amounts of it. Kahn later observed that it's hard enough to deal with environmental issues when we recognize them, but there's not even a desire to act if we cannot remember the way things once were or recognize how bad they've gotten. He's called generational amnesia "one of the most pressing psychological problems of our lifetime."

Face-to-face with night bloomers, realizing that our peculiar, lights-out evening was, for generations before us, just a regular Saturday, Amy's starting to think that generational amnesia is especially relevant to darkness. We've almost wholly acclimated to night as a place of artificial light, and we have generally forgotten the necessary pleasures of the dark. Night, like so many things she cares about, is an heirloom under threat.

"We get used to lights, and we forget that, at one time, people just had oil lanterns. We forget that, before that, people just had fire and candles." Not so long ago and not in some far-off place, but, right here, in the memory of people she knows, people who grew up in and around Honeysuckle House. It's one of the reasons she loves living in a house that doubles as historic artifact—it comes with stories that give perspective. Every bedroom fireplace is a reminder of how her later-installed central heat is a marvel.

It seems particularly poignant that my search for night bloomers has led to tobacco plants, which have their own worrisome form of generational amnesia. According to the *Journal of the Royal Society of Medicine*, the tobacco genus "has probably been responsible for more deaths than any other herb," going on to call it "the most important avoidable cause of pre-mature death and disease in the world."

These are things I pretty much knew, given public health warnings and labels that became legal obligations in my lifetime. But, in the arc of the human story, tobacco has only recently been cast as a villain. In fact, it was once—and for a very long time—considered a creature with life-affirming powers all its own.

Indigenous communities in the Andes have cultivated tobacco since prehistory, and for generations, many cultures across the Americas have fostered diverse sacred, ceremonial, religious, and medicinal relationships with the plant. Tobacco has long been used as an offering that honors relationships between physical and spiritual worlds in a variety of disparate traditions, and it plays a role in some creation stories, which depict tobacco as among the first plants that humans were able to grow on this planet.

To aid public health, organizations like the National Native

Network are working to reclaim traditional relationships to tobacco, which—unlike commercial tobacco use—rarely involve inhaling smoke. Their tagline: Keep It Sacred. The group's efforts to revive generational memory include the encouragement of garden-growing native North American *Nicotiana rustica* as a way of replacing notions of tobacco as something cut and dried with understandings of it as a living being, thirsty and alive.

It might seem impossible to change the course of generational amnesia about anything. But, bit by bit, maybe it isn't. After all—in a reversal of my lifetime of being acculturated to view tobacco as a commercial product—I know that, after tonight, I will never again hear the word "tobacco" without bypassing cigarettes to envision this fragrant garden. Here, tobacco is shelter and nourishment for various creatures, including hornworms and sphinx moths, and I have directly witnessed tobacco's role in sustaining life rather than snuffing it out. It is not unlike how I have been changing the way I connect with natural night, fumbling into ancestral times that, before holding counsel with fireflies, I'd never thought much about.

In recent years, astronomers have started to warn that even the satellites sent up to share human-derived information are now creating substantial light pollution from above, reflecting sunlight with varied brightness. They are basically fake moons, and large numbers of them are being deployed by for-profit industry without much scrutiny, creating light interference that makes it difficult for astronomers to see beyond our species' influence.

Some astronomers, concerned about the increase of visible light from Earth's surface and from low-orbit satellites, have started likening light-polluting industries as relatable to "Big Pharma" and "Big Tobacco" and "Big Oil," going so far as to coin light-polluting industries as "Big Light." NASA reports that, already, there's so much light pollution that they have lost the ability to properly warn against incoming asteroids.

"We talk about things like this all the time in my class on evolutionary psychology," Amy says. "Humans take something that we're drawn to, something that has some positive associations, maybe due to some evolutionary or physiological premise that has made it helpful in small doses, or used in a certain way, and we go on to turn it into something that causes harm. Human-made light, tobacco—there are tons of examples. Figuring out how to stop doing that is one of the great human questions." Because, in too large a dose, any medicine can turn into poison.

Accessing the health of natural night is difficult because artificial light invades other people's space like secondhand smoke. It is, in the shared airspace of this planet, pretty much impossible to escape. Everywhere is, in terms of light pollution, like sitting in the smoking section. People light up and the effects spread.

Even here, miles outside of our rural town—on this turn-of-the-century homestead, where there are no localized lights buzzing—skyglow is present as a sickly green halo in the distance. Amy calls light pollution from the direction of town "Booneshine." It's a locally familiar, cheeky ode to moonshine liquor, which mountaineers famously brewed at night to evade detection during Prohibition. For now, moonshine—that is to say, the reflection of sunlight off the surface of the moon—remains brighter than Booneshine on full-moon nights like this. But, incrementally, it's becoming a competition.

As if just thinking about it makes her crave base-level connection, Amy reaches out to break off a small piece of tobacco leaf, touching the broken plant membrane to her tongue. "Bitter," she says.

"Think that's because of the nicotine?" I ask.

"Maybe."

My hand hovers over bunches of silvery sage until I find the stickiness of a tobacco leaf. I break it open and touch its green to my tongue's pink. The leaf delivers zest that makes me throw my head back, like I'm swallowing a pill, on instinct.

Amy laughs. "I don't know about you, but I'm going to need something to get that taste out of my mouth," she says, retrieving a stash of chocolate stored in a glass jar. "Cacao—now, there's a complicated plant story."

She's referring to the tree that produces the seed that is used to make chocolate, a crop that has been responsible for a great deal of deforestation all over the world—as has tobacco farming—due to insatiable, cross-cultural human appetites for it. Ever mindful, Amy is serving chocolate that has been shade-grown, in alignment with traditional Mayan and Aztec cultivation methods that utilized dark sinkholes, where flowering cacao trees were regarded as gifts of the gods. The revival of shade-grown cacao among foodies—a happy if small-scale reversal of the chocolate industry—has led to increased biodiversity, supporting birds and mammals and reptiles the world over.

Around us, star-seed tobacco flowers, caught in the uptick of a breeze, start to twinkle. Their movement stirs memory of something that Amy has long known but somehow never registered as important until now: Every chocolate birthday cake we've ever consumed, every candy bar we've ever enjoyed, the very confection that we are eating, they have all originated from cacao trees—plants that only bloom at night, leisurely opening at twilight, just like tobacco plants. We are savoring the flavor of trees tended by darkness.

Cacao is cultivated in fifty countries around the world—all of it grown in moon-garden groves that depend, in part, on the cycles of natural night. Amy reaches out to hand me another piece of chocolate. I slip the bittersweet wafer into my mouth, and we sit there for a long while—nursing night's imported confection, basking in the perfume that local darkness is distilling.

Secret Gardens

After my time at Honeysuckle House, I start to suspect that there might be nocturnal wildflowers in my neighborhood. I amass a small collection of field guides and gardening books, including an out-of-print title that was inspired by flowers blooming in containers on a New York City balcony. The book assures me that, for night-blooming wonder, preelectric homesteads are not a requirement.

I start to pay closer attention to flora, night and day, just as, in my search for foxfire, I began to familiarize myself with all kinds of fungi. It seems a stretch that I'm going to come across a night bloomer with my casual study. But, given all that I've unearthed in darkness, I remain hopeful.

Years ago, I kept a vegetable garden by the river, until I decided that, because I am a forgetful weeder and water transporter, it would be best to just let the patch of land go. In other words, it became free to do what it wanted—and what it wanted was to grow daisies and black-eyed Susans and blackberry bramble. Now, the garden is a tangled mass of petal-faces strange and familiar.

Newly empowered, I examine every wildflower I can find, inspect every dab of color. One day, climbing from the river to my house, I'm drawn by bits of yellow along a retaining wall, where

I discover a sundrop flower. It's a day bloomer, but it's related to the evening primrose that inspired Kim, my milkweed-farming friend from Mothapalooza, to start moth watching.

I sit on a stone, warm with sunlight, to study another slender stalk with nondescript leaves. Seen in juxtaposition to the sundrop, I can tell it's an altogether different species. Its blooms are not yet mature. They're held too tightly for identification via my cell phone app. I take a photo and send it to knowledgeable friends. Back inside, I pull out my books. There is, it seems, a chance that I have a stand of primrose growing not ten feet from my back door.

That evening, I intend to take a night walk to see if my suspicions prove correct. But there is schoolwork to oversee. There are meals to make. I forget.

At dawn, after the hustle of seeing Archer off to school, I recall that, if visited in the early morning hours, night bloomers can sometimes be caught before they drop their petals. If my yard hosts evening primrose, it will take a few hours for the sun to burn off their beauty.

I don't have to go far, just past a dwindled woodpile, where I find that some puckered green buds have expanded into tiny yellow flowers that look as luscious as hibiscus. I have not watched them expand in fading light, but I'm nearly positive that this is a stand of primrose. And, to confirm, I will unquestionably be attending their next soiree.

~~~~~~~~~~

At sunset, I confirm the unbelievable: Though I've never planted a moon garden, I've been living in the middle of one. Once I can identify primrose in daylight, I find that my yard is full of it.

In addition to the plants by my woodpile, there are whole fields of primrose downhill from my house. At dusk, they turn buttercup yellow. And I want to revel in them like the bumblebee that, just yesterday, I watched writhing at the center of a shag-carpet thistle.

For weeks, I amble meadows at dusk. In time, I develop a favorite patch of primrose. The stem bits that serve as latches for holding primrose blooms tight release around eight or nine. This is the same time that a family of deer regularly makes its way through the field below my house. The first night our paths cross, the deer come up behind me, making me jump with their annoyed snorts. But on evenings after that, with distance, we find a rhythm to sharing the space, rotating with varied paths on the property—which was, according to neighborhood elders, forest, cabbage field, and then horse-turned-cow pasture before becoming my yard.

One night, after the deer have passed, Wilder trots along after me. Fog gathers on my jacket as we walk through fallen clouds. The ground is damp. Still, I lie on it, pressing high grass into the shape of my body. When I rise, I will leave an imprint, just like the deer who bed down here.

I watch primrose as intently as I watched jasmine tobacco, waiting for the moment when they will drop their latches. When it happens, they turn into tiny ballerinas, skirts swishing. Wilder, frisky though he might be, has taken to lying with me in the primrose patch. Tonight, he remains alert, but, like me, he seems sleepy. We're deep in a river-sung lullaby when suddenly, an unidentified object. Sky, smeared like an out-of-focus photograph. Then, again. This time, Wilder jumps up to make a lunge. I follow his lead, trying to get a better look. A hawk moth.

The creature appears to float over flowers of interest. In some species, moth wings move at 85 beats per second. Watching this still feels less frenetic than doomscrolling screens. I stand statue-still in the moon garden that I've unwittingly shaped by doing nothing at all. This is a moth's true home—in the dark, nuzzling flowers.

The moth is close enough for me to see a straw tongue unfurling to connect with the interior of these night bloomers, which are, alongside *Nicotiana*, among their favorites. The hawk moth and the primrose are puzzle-pieced together, just as I fit into my makeshift nest of sedge grass. This scene is a romantic-era painting, and I've stepped right inside of it, surrounded by wildflowers that, freshly opened, are overflowing with starlight.

The hawk moth flies toward me, then over me. I grasp Wilder's collar. He's prone to chasing butterflies, but I have seen him pause to watch fireflies. The hawk moth seems to have stunned him into curiosity. It's unlikely that the hawk moth will return to this spot once it's gone. There are, after all, other stands of primrose, though the spot I've chosen does feel like a floral smorgasbord. I sniff the blooms that have been abandoned and get a hint of meringue. I consider going inside, but Wilder doesn't pull to leave, so I don't tug, either. And I'm soon glad we've stayed.

When the moth returns, it brings another—and another, and another—until I am surrounded by prehistoric moths the size of birds. They are encircling me, swarming. This moon garden isn't mine, it's ours. The hawk moths come and go to and from places unknown, sometimes navigating mere inches away, where I can hear their scaled wings, fragile and faster than wailing helicopter blades.

Hawk moth. Sphinx. The words feel strange in my mouth

when I whisper them to myself. In Egyptian mythology, the Sphinx was a winged creature that guarded sacred locales, and beloved Appalachian poet Wendell Berry has written that "there are no unsacred places; there are only sacred places and desecrated places." But even desecrated places can be reclaimed.

Native plants have gained popularity in recent years. So has the concept of rewilding, which is generally used to refer to returning a piece of land to ecological complexity. Restoring natural darkness is part of that. But rewilding is a term that can also be used to refer to a person. And, though wilderness is often defined as a place that's uncultivated or abandoned, I think a more apt definition would be a measure of how rich something is in terms of living connection. Wilderness is a place that requires the animals inhabiting it to rise into their senses, alongside other species, in orchestration. And I am wilder for being here, with Wilder, surrounded by these nocturnal beings.

Cyclical darkness, which we all evolved with, fosters interconnectivity that we're only beginning to grasp. This is a moon garden that no human has tended, yet it has been well maintained. Because the primrose knew what to do. Hawk moths, too. I'm enjoying a moon garden that they have been tending for longer than *Homo sapiens* have resided here. Some humans might have generational amnesia, but some other species do not. We are surrounded by land with memories longer than our own. These flowers are of lineages that were growing before I was conceived. They will, hopefully, outlast me with their new-moon faces, in perpetuity.

Scientists at the University of Arizona—focused on primrose species that prefer arid environments, of which there are many—found that the energetic cost of hovering is, for moths, one hundred times that of a moth resting. As soon as the sun sets, they frenzy to find nectar, but they must be careful about how they expend their energy.

Hawk moths have been drawn into tonight's rendezvous not only by sight or scent, but by the rise of the flowers' breath. Recently, scientists have learned that moths are able to sense levels of humidity around flowers in addition to being attracted by color and shape. High humidity indicates that a given bloom is worthy of attention. It's a read of sensory clues that can mean the difference between life and death. Still, I was surprised to learn that nectar is mostly odorless. A flower's fragrance is typically emitted from petals, like pheromones from skin.

These creatures aren't just seeing nectar; they can feel it sure as steam rising from hot, honeyed tea. And, in just a few hours, the blooms we're all enjoying in this wild-sown moon garden will wilt and fall like autumn leaves. But there is solace in knowing that, tomorrow, fresh blooms will twirl into being.

From the other side of the county, Amy continues to send kitchen-garden updates, including word that she recently watched a hornworm pupate, burying itself in a soil-tomb with a promise to rise in spring. "Now that I know he's there, I want to do what I can to help him," she tells me.

Her scythe might call to mind the Grim Reaper, but she's been thinking that her aeration roller, with its soil-needling spikes, is more dangerous. She is already considering how she'll have to be careful when she goes to roll it across the ground next year, lest she accidentally hurt a sphinx in the making.

In addition to planning next year's night bloomers, she's decided to cultivate persimmon trees. She's long thought of growing them so that she might make preserves from their sugar-bomb fruit. But she's ultimately been spurred into motion because they are favored by luna moths. In future years, those native trees will siren-call luna moths' pale green wings into her kitchen garden for communion.

Already, Honeysuckle House is full of old seed catalogs with night bloomers circled. Already, in daily surveys, I'm making conscious decisions about what not to do, scouting mowed portions of my yard that might best be left for wind and birds to sow.

Around us, autumn moves on, colors ebbing and flowing with tides of darkness and light. There are cooler temperatures and longer nights. It's nearing winter. But around Boone, things haven't been settling down. If anything, traffic is amping up.

Tourists have started to arrive in droves, hoping to catch long-range views of fall foliage. In these mountains, we take the leaf-peeping season so seriously that local scientists run calculations about quality of red and gold and the timing of leaf die-off. Telecasters give color reports from newsrooms far off the mountain. There are websites dedicated to reviewing color change at certain elevations. Predictions account for temperature and rainfall. But natural darkness, which is orchestrating the entire show, is rarely talked about.

Now that I've been introduced to night bloomers, it's easier to see all plants as photosensitive creatures. Just as night bloomers' photoreceptors tell them when to bloom in fading light, longer periods of darkness tell trees when to drop their leaves. This has always been a reliable process, a universal language that all sorts of species read. But it's no longer one that we can take for granted.

Over a decade ago, scientists in the United Kingdom started a long-term study to see if artificial light might affect the budding and blooming of urban trees. They were surprised to find that spring budding and autumnal leaf-fall were altered in communities of high artificial light by more than a week.

At first, they thought it might be due to heat pockets caused by urban living, or some other temperature-related symptom of climate change. But evidence ultimately showed that the shift in seasons was solidly due to photosensitive reactions. Artificial light, all on its own, had become concentrated enough to shift seasons. Like moths that forget to pupate due to light, under duress, trees forget to pull nutrients back into their bodies for winter dormancy. They forget to let go of their leaves. They cannot sort the temporal-cue difference between natural light and the light humans create.

I'm beginning to think that, in a time when the Milky Way is obscured by shrouds of our own making, we might already have a level of generational amnesia that renders stargazing a nonfunctional enticement. We have inherited, across cultures, sacred-sky stories. But we generally can no longer see the starlight that our ancestors were interpreting.

We cannot, as individuals, meaningfully reduce light pollution enough to revive massive darkness; we cannot easily solve systemic problems. But we can change our immediate environs in ways powerful enough to alter the health of nearby night bloomers and the trees we know personally. We can take care of ourselves, and each other, by darkening our yards and neighborhoods, starting immediately.

Maybe, if actions were taken in enough places, by enough night-curious people—through individual actions and increasing demand for regulations that address the public expense of artificial light at the local, regional, national, and global level—natural darkness could start to knit itself back together like bone and tissue, like living nets of mycelia.

At the peak of leaf-peeping season, Amy invites me back to Honeysuckle House for a potluck to mark the shift to greater periods of darkness. On my way to the merriment of hand-cranked apple cider and picnic tables covered in casseroles, a stoplight forces me to pause just outside of Boone, right next to the sign that welcomes visitors—many of whom come to enjoy the Blue Ridge Parkway, a famed scenic road that runs through our county.

The sign features wildflowers and the wood-carved likeness of an autumnal leaf. Adjacent highway edges are usually trimmed by road crews. But this patch has escaped the attention of weed eaters.

Alone in my car, I start laughing. I cannot help myself. The profile is unmistakable: Slender green stalk. Tight leaves. Pursed buds. There is a substantial stand of primrose here, frozen in sun-retreat stance, poised to greet leaf-seeking pilgrims. In these mountains, where visitors have come to watch dead leaves fall in daylight, there will be blooms rising, mostly unnoticed, in darkness.

Undoubtedly, many of the leaf-lookers have stands of primrose growing in their own neighborhoods, along their own municipalities' unkempt curbs, state beyond state. There are more than 400 species of the *Primula* genus across the Northern Hemisphere, with more than a few night bloomers. Many are not, by human hands, tended. Moon gardens, particularly those of the wild variety, are unique to their own dark ecosystems. But if we go looking for them, pretty much anywhere moths roam, we will find them, diverse as their native environs.

Nocturnal-blooming flowers grow in both temperate and arid environments. In swamps, deserts, prairies, rainforests, suburbs, and city parks. Everywhere, there are secret gardens hiding in plain sight, under the cover of daylight. Still, I know from experience that it takes practice to see the wildflowers for the weeds. Because experiencing their charm requires that we allow our skin, alongside petals, to be steeped in darkness and doused with moonshine, unadulterated.

Humans Surviving

~~~~~~

Burnout

I've been with wilderness survival instructor Luke McLaughlin for less than ten minutes when he asks: "Would you like to practice having a mini death?" The invitation catches me off guard. Luke notices. "By that," he says, "I just mean, let's sit and be still and quiet for a few minutes."

Silence. Stillness. They are experiences of reduced energy. He's suggesting that we should go dark for a minute. And I can see what he's getting at: Reduced energy is, sometimes, what we need to go on abundantly living.

I've never been good at pretending. So just now, when he asked me how I was—after welcoming me to his field-and-forest acreage, located not far from Celo—I did not give the platitudes of convention. And, given that I answered honestly, I suppose I shouldn't be surprised by his suggestion that we should pause for a minute. The briefing wasn't pretty.

I have, over the past few weeks, not been sleeping well. I have also been consuming too much caffeine. I'm having trouble processing the massive ailments of the world on top of my family members' needs, elderly to preteen. It's been a particularly hard couple of days. I've been trying to support Archer, who is, as a twelve-year-old, being asked to do homework that requires having ten open computer tabs. His school system no longer issues books. No child in this county is required to feel the sensation of wood pulp under their fingers as they turn pages with necessary tenderness. Instead, their tactile reality is plastic keyboards that invite banging, white-hot screens that they're rarely allowed to look away from.

All around me, people seem to constantly be talking about catching up, but catching up to what? I cannot seem to even keep up. Everything's moving too fast. It's too much. My mental load is tinder, set ablaze. Burnt out, that's what I am. And yet, in what seems the antithesis of what I'm needing, I have come to Luke for a lesson in primitive fire-making. Despite my charred brain, in the process of attempting to reconcile with natural darkness, I'm seeking to build open flames. It doesn't make sense, not even to me.

I wish I could say that my dips into natural night have proven to be a cure-all for thriving in a harrowing age. But though they have provided solace and insight, I still belong to a culture ruled by electronic technology. I've become more comfortable roaming around outside in the dark. But, indoors, I'm still drowning in artificial light.

In 2020, researchers from Monash University in Australia conducted a study that required participants to wear devices measuring their artificial light exposure at night the way nuclear plant workers wear badges to alert them to radiation exposure. In 50 percent of the homes studied, light levels were high enough to cut people's sleep-inducing melatonin hormones in half. I think my household would have been among them. Going dark indoors still isn't my forte. Then again, my species is one that's always excelled at firelight making, whereas the ability to lower light pollution is eluding us—maybe because, until now, it wasn't a skill we needed.

Ever since I spent time in Amy's lush, history-inspired moon garden, I've been thinking about the lineage of artificial light in the human story. That's what originally gave me the impulse to trace artificial light back to its beginnings. Firelight has shaped the human relationship to darkness, and studies show that it doesn't disrupt human physiology the way modern artificial light does. Neuroscientist Randy Nelson's research on how deeply artificial light disrupts our circadian rhythm—which regulates not only sleep, but also digestion, and body temperature, among many other bodily systems—has led him to suggest that though it's the "paleo diet" that gets attention, we should be focused on "paleo darkness." But what does that even mean?

Up until, well, pretty much yesterday, humans lived in small groups and spent their nights under the influence of handmade firelight. To go paleo-dark is to know darkness fireside. Low lighting is what we of the modern world tend to call ambience—often reserved for special occasions. But, for almost all our history, ambient light at night was the norm. Now, every enclosed space seems to carry the blare of big-box stores and casinos, which purposefully use lighting schemes to encourage careless decision-making among humans who are gambling.

Scientists have found that, for people who lived with preindustrial darkness, natural night was, in part, a waking experience, and I want to find out what I've been missing. I've enlisted Luke's help because—akin to the way Amy likes to linger in the Victorian era—Luke is a part-time resident of the Paleolithic with a penchant for primitive technology. And fire-making is, by far, his most requested lesson. I first met him years ago at a primitive-skills gathering run by mutual acquaintances. There, he taught me how to use an atlatl, a projectile weapon that was a precursor to the bow and arrow.

My paternal grandfather was a target archer. My son is named in homage to him. I always thought I might visit Luke to build a bow for shooting, but, instead, I've come to learn how to use the primitive fire-making tool of a bow drill, which consists of the same bent wood and string that a standard shooting bow does, only utilized in a way that helps humans produce flames through friction.

Luke—who has a life-size raven tattooed on his chest and blue-ink antlers that curve around his neck—is ever mindful of where he focuses his energy, because, as a survivalist and hunter, he knows the true cost of replenishing what he spends. Right now, by necessity, he's using most of his measured attention to settle me. I've come to handle fire-making technology. But, first, I've got to get ahold of myself.

"Firelight is a stimulant," he says, "so we need to make sure we're ready for that." He directs my attention to an oak tree at the corner of his land. "You said everything's moving too fast. What about that oak tree? Is it moving too fast?"

Okay, okay. I know. But, silently, I'm already protesting: A tree doesn't need a computer science degree to parent a preteen! A tree isn't asked to come up with a new password for every daily activity! Recently, a friend told me that she downloaded an app to remind her to drink water, and I didn't even think it was all that strange—helpful, maybe. This is where we are: losing our senses to the point that we're beginning to outsource our cues to hydrate.

But would a tree forget to drink? It's not that far-fetched. After all, if this oak was under a streetlight, it might forget to drop its leaves. When a caterpillar is exposed to artificial light, it doesn't always remember to pupate. Every living being on this planet seems to be on the brink of being overloaded with energy it doesn't know what to do with. None of us—not plants, not animals—are alone in this.

"How are you feeling right now, in your body?" Luke asks.

"I feel like I'm full of static," I say.

Static is light-dark-light-dark oscillating too quickly. It is its own sort of signaling, which anyone who has ever lived with an old-school, rabbit-eared television recognizes as: *Connection lost. This is not working.*

We walk past Luke's rough-hewn barn to a picnic table littered with found objects: a sun-bleached possum jaw, an acorn. Luke does not speak for several minutes. He watches trees sway. I watch trees sway. The inaction makes me, restless member of a restless species, almost squeamish. I shift in my seat. A lot. Then, a little.

Luke's massive dog, Fenrir, a black shepherd with pointed wolf ears, walks by. My hand grazes his coat, dark fur rippling under my fingertips. In the quiet space Luke has created, I start to feel my breath slow. The static in my mind begins to melt like snow overpowered by rain.

I can hear the trickle of a nearby creek. Shriveled leaves that have not yet fallen rattle above us. I did not expect fire-making to begin with a lesson in mindfulness, but most primitive fire-making instruction starts with an ecosystem introduction. Because, for our ancestors, wherever they roamed, light-making required ground-level intimacy with other species.

Finally, Luke speaks. "Let's orient to the land you're going to know for fire-making."

He directs my attention to the spring-fed creek as a safe place to drink. He points out his house on the hill, where he lives with his wife, Luna. I express gratitude for this slowing. It is, he says, his natural tendency. "My energy runs around the speed of moss and rocks," he says.

"Mine runs more the speed of a hummingbird," I say. "Hawk moth, maybe."

"It takes all kinds of energy to form a functional community," he tells me.

It's a generous comment that inspires me to admit that I'm confused by my impulse to make fire, since, even on a good day, I tend to be a live wire. But Luke assures me that, counterintuitive as it seems, he thinks I'm onto something, given my quest to better relate to natural darkness. "We only know warmth by knowing cold. We only really know life by acknowledging death. We only know darkness by understanding light," Luke says. "It's like yin and yang."

When I was around Archer's age, I wore a yin-yang necklace. I vaguely understood its black-and-white swirl design to be symbolic of balance. But recently, I discovered that yin and yang are, quite literally, mountains. Yin translates as: shady side of a mountain. Yang is the sunny side, a south-sloping hill.

In early Chinese cosmology, mirrors were sometimes placed on the yang side of a mountain for concentrating sunlight to start fires. And there were mirrors placed on the yin side to collect the dew that could put those fires out. Day was the ultimate yang. Night, the premier yin. What one shrouds, the other reveals. Without each, the whole cannot be known. Without both, there cannot be balance.

I have been walking the yin and yang of Southern Appalachia. But even now that I'm acquainted with both sides of these mountains, I'm not sure how to balance artificial light and natural darkness in my nightly life, within myself. That, as much as anything, is what has brought me here to time travel.

"You know, when people tell me they're grateful for something, I can usually see pain underneath," Luke says. "If they say, 'I'm so happy to be connected to community,' it indicates that they've spent time feeling disconnected. If you're grateful for darkness and the biodiversity it supports, to me, that means it's something you've been deprived of knowing. When you start pulling on one thread, you see that it's connected to others. When you start loving and appreciating something about the environment around you, you realize the other side of that love is grief over what's happening on a global level. It's something I see in people all the time as they gain a greater awareness of nature."

It's true. As I've embraced natural night, I've had to simultaneously grieve it. There is no marvel under its protection that is not threatened. Darkness itself is in danger. And after spending several seasons in the company of fireflies and glowworms and foxfire, I need to meet myself as a fire-making animal. I want to understand why, after all my dark discoveries, I'm still struggling to let go of so many lights in my life.

Luke has asked me to bring a few tools, which I lay on the picnic table. They're things I had to hunt and gather at a local general store that still sells hardware alongside barrels of peach candy and gift-packaged grits. In addition to the hatchet he suggested, I've brought a hand drill. In response to the knife request, I've chosen a fixed blade.

"These are things I would've suggested," Luke says, giving me a high five. I take the gesture to mean that I've been cleared for exposure to firelight. He has, from the start, been considering my inner-outer energy balance.

He puts down the knife he's admiring and says, "So, if you're trying to make an ember, a flame baby, what would you need to have it catch fire?"

"Tree fibers?" I know this only because I've heard that poplar bark is a prized fire starter.

Luke nods. "Just like people, trees have different energy levels and survival strategies," he says. "They have different techniques to fight entropy. Locust is dense, so it doesn't break easily, which means it doesn't allow bugs and fungus in much. Pine and birch use resin to resist. They give us what we need for fire just by taking care of themselves."

He directs my attention toward the woods, where almost all the trees have dropped their leaves. I've come to think of leaf litter as biologic fairy dust that feeds the most wondrous forms of nocturnal life, yet common terminology refers to it as trash. "Do you know another word for leaf litter?" I ask.

"Humus," he says. "That's one I've always liked because it's the root of the word 'humble.' It's a reminder that I'm of the earth. I'm of this soil. 'Humus' is a grounding word."

Core Strength

We enter woodlands with the creek gurgling at our backs. Even though I know it would be helpful to have poplar bark for fire starting, I don't know how to find it, because I don't know how to read the nuance of bark. In a warmer season, I would recognize the trees' flowers, rubbery blossoms befitting this temperate rainforest. As a child, I would push their magnolia-thick petals apart with my fingers to find bright-orange markings at their center, each petal playing a role in making their circular designs whole.

Even if the trees had leaves, I would be able to identify them. But on the ground, they have already disintegrated, leaving little trace of their former selves. Luke points out the species' arrow-straight trunks so that I might recognize them without their outerwear. When I understand the way their branches split toward the top of the canopy, I see them everywhere.

He urges me to look for fodder, but the ground is damp. Dry tinder doesn't seem promising. "Making fire is learning how to live in the cycles of ecosystems. There's so much moisture in this region, finding dry branches can be hard. Where do you think we should look?"

We're already on the south-facing side of this mountain. I'm not sure how else to orient. Then, Luke points up, where limbs form a net above us, catching deadfall before it hits damp soil. There's a wealth of what we need there, caught between earth and sky, where, by wind and sunlight, it's been dried.

I untangle loose twigs from branches. "Remember, we're not just looking for poplar," Luke says. "This isn't like going into the grocery store to grab something specific, like a jar of peanut butter. We're going down every aisle to see what we come across, looking for whatever might be useful."

He extends a twig so that I might catch a whiff. "Aromatic," he says. It has the delectable scent of something that should be topped with maple syrup. "Look at the bumps on the side—what do you think this is? People use it as spice. That's your last clue."

"Spicebush?" I guess. Luke smiles. I've brought just enough language to start naming the world.

"When something smells, it's an indication that it holds volatile oils that are often good for fire-making. Smells are those oils turning into vapors," he says. "It's an aromatic rising of carbon."

Luke notices a locust post with sable-colored fungus growing out of its north side. A piece of it has fallen. He picks it up and knocks on it as Celts once knocked on wood to summon a tree's protective spirit. "Tinder fungus," he says, "If you light this, it will smolder for hours."

Before humans learned how to make fire, we learned how to manage fires started by lightning. In some cultures, people spent whole lifetimes guarding and moving with bits of a single hearth. In time, humans mastered combustion, but fire was still valuable. It was valued. "To me, fire always has an element of ceremony," Luke says. "There's a reason we carry a torch at the Olympics. An everlasting flame is a common concept. Our human ancestors, in some parts of the world, they passed a single fire from one generation to the next."

It was the ability to manage, make, and transport fire that gave our species freedom of movement. It's how we've been able to make homes in blizzard-prone places, from Appalachia to the Alps. When humans migrated, this slow-smoldering fungus was a tangible bit of home terrain that could be carried in hand. Luke passes the fungus to me. It weighs almost nothing.

Tinder fungus was found among the belongings of Europe's oldest-known mummy, the Ice Man, who was walking this world in the 3000s BCE. His body was discovered in mountains between Austria and Italy. Packaged with his fire-making tools, archeologists found pieces of fungus resembling the golden half-disc I'm holding. This fungus is not bioluminescent. But in human hands, it learned to glow nonetheless.

~~~~~~~~~

Luke is walking through the woods barefoot, unbothered by rugged terrain. When we reach an area guarded by bramble, he pulls thorns back so that I can pass. "I call these awareness enhancers," he says.

Deeper in woodland, he lifts the limb of an evergreen tree for my inspection. "Know what this is?" Hemlock. We both recognize it as grief. Once, hemlocks ruled Southern Appalachian forests. Now, to see one is to know that woolly adelgids, invasive insects that topple giants, are coming for it. "We know it will have a shorter life than its ancestors, but some still get pretty tall," Luke says. "Hemlock's one of the best for fires."

There are dead branches toward the bottom of the tree.

They snap between his fingers easily. "Building a fire is about learning how to integrate into a landscape," he says. Luke tucks more branches into his bouquet. I gather birch bark from the understory, silvery curls so oil-saturated they seem to shine. "Sometimes, when I'm building a fire, I feel like a chef, and like any chef who has good ingredients, I try to let them speak for themselves by letting the character of certain trees come through. They all have different strengths," Luke says. He moves on, toes curling around stone. "Making fire is like following a recipe that draws from skills and awareness. If you're a human who doesn't know how to make fire, it's like you're a monkey who's forgotten how to climb a tree."

There are theories that suggest the human ability to build fires is a factor that convinced our ancestors—who once slept in trees, not unlike owls, and moths, and bats—to come down to rest in ground nests, the origin of beds. Fire is what allowed us to shed our fur and run faster than the animals we were hunting, since we could expel our body heat and gain it back through fire, outsourcing the regeneration of corporeal warmth.

Through cooking, we similarly outsourced the processing and digestion of our food, letting fire do the preliminary work of breaking things down, which ultimately changed the size of our teeth and the shape of our jaws. It was also around the time that we started using fires for cooking that our ancestors' brains began to expand. Fire, as an organic extension of our bodies, shaped us into modern humans. And, through fire, modern humans keep shaping the world.

Dating as far back as 92,000 years ago, our ancestors used fire to alter ecosystems. In North America, the land European settlers viewed as wilderness had already been greatly altered by human relationships to fire. In California, where wildfire is an extreme threat, Karuk and Yurok communities have been prescribing human-created fire for more than 13,000 years, making small burns to prevent large disasters. But historically, the U.S. government passed laws to sever those human-fire relationships, in part because they have spiritual significance that settler-colonial culture denigrated. In recent years, the Forest Service has been working with tribal representatives to reintegrate land and culture, attempting to heal ecosystems by honoring traditional knowledge.

Fire historian Stephen Pyne has proposed that we've already entered the Pyrocene, a geologic period in which the human use of fire has come to a head, because humans have, in general, lost their working, ancestral relationships to it. He calls the Pyrocene "the fire-informed equivalent of an ice age" and suggests that humans' "primary ecological signature is our ability to manipulate fire." Only, with each electrified generation, it's gotten harder for us to recognize that we're playing with it at all.

Heat and light were, in the event of internal combustion, separated. We can no longer touch the stove to learn that it is hot. LED light does not, from its luminous point-source, emit heat that would act as a tangible warning. The enclosed realities of a lightbulb with an on-off switch gives our bodies no real tactile clues to the meltdown forces we're handling. "With increasing frenzy," Pyne writes, "humans are binge-burning fossil fuels . . . Combustion acts as an enabler."

We want more light. We want more everything. It's hard not to, because we are an exploratory species—and we've created an environment that gives us little ecological constraint, so we follow, follow our evolutionary instincts without direct, larger-world feedback. But, out here, things seem different.

I feel a tiny explosion on my arm. Then another. The sky darkens. I guard the tinder I'm carrying, and my skin is covered in rain splatter. "Are we still going to be able to make a fire?" I ask Luke.

"Hope so. It's just spitting on us," he says. "Rain isn't a deal-breaker. It just makes getting and keeping a fire harder."

When pine trees go down, their wood gets soft. But the places where their trunks bulge don't break down easily. Luke points to a fallen pine with joints swollen as my late grandfather's three-times-broken elbow. "The trees put more resin in those spots so they're resistant to rotting," he says.

Luke lifts a splintered piece of wood from the ground. Scent, forest green. "God, how I love that smell," he says. "Sometimes, I burn this like incense."

He directs my attention to one of the tree's bulkiest parts. "People call these sections fat lighters. They're one of the treasures of pine trees." They're not so much for starting fires as for sustaining them.

I study the still-standing pines around us, their knotty sections familiar. It's like gaining X-ray vision into the interior of pines in my own yard—golden rings, held safe in all their cores.

Luke traces contours of the fallen log. Some parts are moss covered and others are webbed with mycelium. He says, "We're going to take one of these fat lighters since it's starting to rain, just in case."

He hands me a small axe. In the shade of evergreens, I chop

until wood loosens. I lift a splinter of resin from the forest floor. Against the gathering clouds, it looks like caramelized sunshine.

~~~~~~~~~~

When Luke asked me how far back I'd like to travel into the history of human fire, I declined flint and steel, as it felt too close to industrialization. A hand drill, which is composed of only a baseboard and a dried flower stalk, sounded a bit too far in the other direction, so I settled on a bow drill. In the bow-drill process, a pencil-like piece of wood, known as a spindle, is twirled around using a bowstring while pressed against a wooden plank. And, wood-against-wood, this creates an ember through friction.

"A bow drill is physically easier than a hand drill, but it's harder mentally," Luke says. The mechanics of the bow outsource part of the brute force required, but juggling the extra parts can get complicated. Physically easier, mentally harder. This seems, to me, the definition of modernity.

The moving parts required of our lives are, increasingly, not mechanisms we're holding in our hands, but the untold number of things we are forced to manage in our brains. So, rather than or in addition to physical strain, we're all experiencing some level of mental hardship. Our lives are all about energy, expended and absorbed, without much tactile recognition that anything is being exchanged. More and more we outsource bodily tasks. We give away our senses. Through artificial intelligence, we've started outsourcing our thinking, as if we're willing to numb ourselves out of existence. For Luke, making fire—and rewilding work, in general—is about reclaiming corporeal awareness.

"When I first got interested in earth skills, fire felt like a gateway for me to connect with ancestors, deep ancestors who came before agriculture. Their lifeline was fire, and I became obsessed with it." He holds out his hands, palms up. "When you use your hands to make fire, you get calluses. When I started practicing with a hand drill, I could have bleeding hands, but I kept going. I asked myself all the time, 'Why am I doing this?'"

It's what I've been asking myself since I decided to act on this light-making impulse. But I suppose it's like questioning why anyone would want to grind wheat to bake their own bread when they could buy a factory-made loaf at the store. Why would artists spend years painting with stone-crushed pigment when they could just use a printer? Why play instruments built from fallen trees when music can be created via computer programs? Because, without tactile processes, there'd be a lack of more-than-human communion.

Luke says, "For me, making fire was like channeling the power of my body." It's what led him to recognize that he had power in the first place. Actual energy to burn. And, once he'd mastered the fire-making process, he quickly moved on to other skill sets.

Daniel Fessler, an evolutionary anthropologist at UCLA, has done research that makes Luke's obsession with fire, and my curiosity, seem less like personality quirks, more like vestigial instincts. Fessler has found that fire-making—a fundamental activity for our species—is a skill that has typically been mastered mid-childhood. In Western cultures where children are not given opportunities to handle fire at young ages, they tend to hold a fascination with it at much older ages. It's been suggested that this might be why people like fireplaces in their homes, even in the presence of electric heat. We crave a primal relationship with fire because we've coevolved with it. We are still, as a species, intertwined with it.

Interestingly, since vegetation for burning varies so greatly around the world, Fessler's research suggests that we have not evolved with specific plants, as hawk moths have with primrose. Instead, we've evolved to carry behavioral patterns that make us adept at concentrating energy with whatever we can get our hands on. In California, the process might require intimacy with boxelder and cedar. In Florida, sable palm frond. In Idaho, a fire-starting human would learn the qualities of sagebrush and cottonwood and yucca. The magic lies in the energetic interplay of humans and specific ecosystems. Light, like the makings of a fine meal, was once sourced locally.

As Luke compulsively worked a hand drill in his early fire-making days, he often found himself chanting to no one in particular, "Please be a fire."

"It took me a while to recognize that mantra as a prayer," Luke says. With almost every early technology that humans used to make fire, the process required moving up and down in a motion that's used in prayer traditions the world over.

It took weeks for Luke's prayers to be answered. It was only when he was too exhausted to disparage himself that pain yielded to woodsmoke. "The most important part about the fire-making process for me, early on, was how it taught me how to make fire without burning myself out."

I feel the sting of a tear that does not fully form. Yes, this.

I might be a novice earth-skills student, but I'm a master of combustion. I start tiny, enclosed fires all the time. I imagine all the cell phones and televisions and computers in my house piled in my yard, logs on a blue-flame fire, concentrated. I've been getting my energy prepackaged. I cook in ovens where there's heat but no light. I turn on overhead bulbs where there's light, but no heat. And because the energetic costs are hidden from my senses, heat and light divided, something gets lost in translation.

Rachel Carson once wrote: "There is something infinitely healing in the repeated refrains of nature—the assurance that dawn comes after night, and spring after winter." In her lifetime, the notion that night might stop coming after day due to human interference was not a pressing issue. The concept that summer might melt into winter, through climate change, remained a fringe one. The idea that losing insects and birds to artificial lights might cause silent nights wasn't something that, in that time of relative darkness, would have been particularly worrisome.

Some moths have already evolved to avoid artificial light. In one study, rural moths from dark areas and ones from light-polluted cities were collected and raised in captivity. When released, it was confirmed that moths descended from those in the heavily lit places avoided artificial light much more than those from naturally darker places. After generations of exposure, city moths were no longer tricked by fake moons. Is this a hopeful sign, or a sad one? In scientific terms, it's simply adaptation.

In almost every situation where extreme artificial light is involved, it makes our sensory lives smaller. Blue light can hurt human eyes in large doses, same as owls. But, with humans, harm tends to be a slower burn.

Recently, a study found that young adults who regularly use electronic devices triple their risk for myopia. It's a global trend and researchers suggest that it was exacerbated by the pandemic, when screens became, for some people, their only portal out of four walls. More and more, humans cannot see at a distance due to the use of handheld screens, which potentially cause damage via light and the close distances at which we've been training our attention. Researchers suggest that by 2050, half of humanity will be shortsighted.

~~~~~~~~~

Luke has an entire fire kit tattooed on his arm, an artful rendering of a spindle stick and a baseboard. It documents some of the actual charred friction-fire holes he's ground down with primitive tools pursuing light. On his knuckles, he balances a constellation of scars, collected in a variety of biomes. He can point to each like a map and tell you where the original wound was formed— many of them while fire-making. "Most primitive-skills people have hands covered in scars like these," he says.

He holds up a sample baseboard that matches the one on his arm, so I can see what he's asking me to make. The wood piece is the shape and size of a large soap bar. This is where the spindle will be braced for friction. "We won't find heat if it's too easy to move the spindle," he says. "Fire is concentrated energy. We have to embrace the hardness of this. The energy we expend to make friction is what makes this process hard. It's also what makes it work."

Luke then shows me an example of a spindle, which looks like a large pencil. "This is going to help us concentrate our energy. It's like the difference between wearing bowling shoes and stilettos. You're walking with the same weight, you're just concentrating it. As you spin this with the bow drill, you're destroying the spindle you've created. You're destroying the base-board you've made. You're putting your energy into an ember that you're then going to put into your nest."

By nest, he means the tinder bundle I've made of poplar bark, coiled under his direction to look like a tiny basket. He has a few basswood branches for me to choose from in making my own baseboard. "Like Michelangelo said of sculpting," Luke says, "you're just going to have to take away the parts that don't belong." He lifts a log from the forest floor to use as a mallet, showing me how to go through a battening process, using the log to tap my machete.

After my first few moves, he steps away. The wood shifts. I shift. "You're doing well," he says. "Nice correction." In my hands, the branch is reshaping. With my mallet, like a sculptor, I keep tap, tapping. Next, I need a spindle. I follow the same process as with my baseboard.

Before long, I'm whittling. Basswood shavings gather until it looks like I'm sitting in a snowdrift. "I think that's enough," Luke says. "Maybe even a little too much."

I hand my work over for inspection. He presses his thumb against the spindle. It snaps. My creation would not have survived the pressure of a bow drill turning it with string. "All that work," he says, shaking his head. "But, sometimes, this is just what happens."

He offers me a spindle from his personal collection. It feels a little like cheating to accept it. But the sun is going down.

One end of the spindle is pressed against the baseboard, where it will gather heat. The other is braced against a socket piece. He twists the spindle into the bowstring with a flip, then he starts moving it like a saw, spindle grinding against the board.

"Once it's smoking, there's always a chance you have an ember," he says.

People with upper body strength just push down on the bow drill, but he's twisted his arm around his leg in a stance he uses to teach children and adults who, like me, lack muscle. To make it less intimidating, Luke suggests that I think of the position like a yoga pose. But I am not a yogi.

Still, I copy his pretzel posture: Spindle in hole. Hand under my kneecap. Foot against board. Bow in my hand, sawing. Forward and back, as if against a stringed instrument.

"Once," Luke says, "I taught bow drill to a cello player. They were like, 'Oh, I know how to do this.'" I am not a cello player. But I am soon holding my bow drill steady.

"You're doing great," he says. "You don't have to do anything other than what you're doing." Smoke begins to rise in wisps delicate as cotton thread. "Keep breathing," Luke says.

The reminder is helpful. Bow forward inhale. Bow back exhale. They are long strokes. I take larger breaths. My bowstring is inching upward. Luke leans in to adjust it as I saw away. "You're almost there!" he says. But my muscles are tired. My energy, dissipating.

I've started to think about the complexity of what I'm doing. The migration of energy from my hands to my mind has caused me to favor my upper body, as if my brain alone could do this. I'm no longer putting my weight into things. I lose contact, going from a woman who has the whole world backing her up to an individual with no grounding. Within seconds, the promise of an ember is lost. My spindle rolls right out of my bowstring.

"Try again?" Luke suggests.

"I'm not very good at working a bow drill," I say apologetically.

"It's almost like you've never done this before," he says. "You're actually doing very well for your first time. But how does it feel, not immediately being good at something?"

"It makes me want to laugh," I say. "That's what I tend to do when my attempts seem ridiculous."

"There are worse ways to react," Luke says.

I try again. Same motion, same breathing pattern. Success.

There's smoke coming from the board. But when Luke moves in to transport the ember to my tinder bundle, it falls to the ground. Soil claims my spent energy—light, buried like a seed.

"I almost never lose an ember," he says. "I think it's because you were meant to do it, but I stepped in. Next time, I'll step back."

I thread the spindle. I flip my bow. But when I start to work, the spindle does not catch on my bowstring. Instead, it launches like a dull-tipped arrow, nearly hitting Luke in the face.

I begin again, but my baseboard groove is spent. I need to make another with my knife's tip. I reach for my blade. But I'm careless. I cut my finger. Crimson wells on my pointer. "Good thing you told me about the scars," I say, cradling my hand. "Makes this feel more like initiation than failure."

"Accidents happen fast," Luke says. "You need to settle your mind. In the meantime, we better find something to slow the bleeding."

His first instinct is to look for a coagulating plant. Mine is to find a plastic Band-Aid. I'm new to deep-time living.

I press my finger against the black shirt I'm wearing and rummage in my store of supplies for a first-aid kit. When I return to my fire-making tasks, fully bandaged, Luke tells me that he's concerned about the fading day. Making fire in darkness is not easy for beginners, especially with an injury. "Fire has been with us for thousands and thousands of years," Luke says. "It will be there for you, if not today, tomorrow. You can keep practicing. I don't want to rush you, but the sun is setting."

He's close to calling this session. My finger throbs, but I do not want to give up. I take my position. I get into my baseboard groove. Slow is steady, steady is fast. But when I'm close to creating light, I lean in. My energy is, again, lost to the wind.

"What happened?" Luke says. "I saw you try to power through there at the end. You were like: 'I'm going to will this into being!'"

"Yeah," I say, "that's my MO."

"Being able to power through is not a bad trait," Luke says. "And I bet it works for a lot of things. But it won't work for this. This is about finesse." It is about a balance of exertion and rest.

I think of Luke bowing with his hand drill, praying all those years ago, his hands bleeding as my hand is now. I feel a dull ache in my torso. "I'm working muscles I didn't even know I had," I say.

"Yeah," he says, "a bow drill works your core."

By necessity, I accept that I might not be able to do what I set out to do. I give myself grace. It's what I've been needing. It feels like what the world has been lacking, everyone and everything seemingly rubbing against each other, bones with no soft tissue between. Friction fires, all over the place.

I release expectation. I promise myself forgiveness, even if I fail. *All I have to do is breathe*, I tell myself. If I get out of the way, my body will do this without me having to think. *All I have to do is breathe*. This chant is, oddly enough, the same one that got me through a twenty-one-hour labor, years ago. The core pain I'm experiencing with this bow has tinges of the ones that brought Archer into being.

The chaos of my mind migrates to my hands. The concept of putting my energy into something has, in this moment, literal meaning. I am supported by the forest behind me. I'm held by the earth below me. I gain confidence. Then my spindle cracks. Luke has another in his fire kit.

"Start again," he says. I'm close. We can both feel it.

The sun is setting. I flip the spindle into string. I cup my bleeding hand around wood and begin again. The spindle's whirl makes music, a hum I have not noticed until now. In and out, I breathe. Back and forth, I bow. Finally, a thread of smoke.

"Keep going," Luke says. "Keep breathing. You're almost there."

In an instant, something shifts between skin and blood and bone. The spinning of my abstract mind fully transfers into the spindle's wooden whirl. But it is hard to sustain. I fall to the ground, exhausted. But Luke's keeping watch. "I think you've got something," he reports.

Soon, it's confirmed: An ember has been born. The smell of woodsmoke, unmistakable.

We stare into a tiny mound of charred wood dust, the leavings of my labor. To give the ember oxygen, Luke tells me to wave my hand over it, as I once did the wings of a dying moth. Then, under Luke's instruction—as I once blew against that cecropia's head—I exhale, gently. In the mound of charred wood dust, a crimson glow appears.

"Yes! You've got an ember!" Luke exclaims.

He hands me the wood-fiber nest spun in preparation and says, "This time, you should do the transfer." I flip my board so that the glowing ember falls into the darkness of my tinder nest.

When my basswood-mingled energy meets poplar bark, the ember glows. The ember grows. Illumination dims and rises, just like the lantern of a firefly. From a distance, it might very well be mistaken for one. I laugh at this confluence, the interplay of all that has led me here, full circle. I have fire in my belly. It is an observable entity. And it looks just like the living lanterns that inspired my night-loving journey.

Luke instructs me to pinch the kindling. "As the ember gets bigger," he says, "you can progressively blow a little harder."

I hold my tinder bundle close and blow. The ember fades and shines. The languid interplay of dark-light-dark-light tells me when and how deeply I need to breathe. I follow the ember's lead. My breath, stronger. Its glow, larger.

Luke has moved into darkness to prepare a hearth site. Nest in hand, I follow the shape of his voice. And, just as I reach him, the nest ignites. "There it goes!" he says.

Flames consume bark and joy bursts through my body. My fingers are protected only by a tuft of tinder. It's time for me to release this inner light turned outer fire, or else my skin will singe.

I put the bundle on the ground. Luke and I sit on opposite sides of the burn. Me to the north. Luke to the south. Between us, fire, a dancing energy mount. "You made this," he says.

It is not only a physical accomplishment. It's something in the realm of spirit, for which there is no spoken language, only the poetry of what's been lived. This is a chemical reaction; it is the divinity of creation. On the ground, fierce petals of yellow bloom from the tips of deadwood twigs. In the air, ghost limbs branch out as smoke, outlines of what this firewood was before. I can taste a hint of spicebush on my tongue, ethereal communion.

Fireflies can make light inside of their bodies—and we can make light outside of ours.

If other animals studied humans the way we study them, this would be our notable feature.

Maybe the wonder we have when we look at fireflies is some recognition that they embody our hidden powers, the fire we all, instinctively, yearn to create soon after we enter this world—each of us, legacy of fire-making ancestors. Every human who exhales has the potential to breathe fire. We are stardust, reshaped into creatures capable of burning whole planets. We know this, practically. But the sensation of enacting this evolutionary behavior has proven powerful beyond imagining.

"This is my fire," I blurt. It is not a declaration of ownership. I am this fire; this fire is me. I recognize my energy in the flames. This light's combustion was not hidden. Its catalyst was me.

I've lived out the cost of human-made light, and I am grateful for the struggle. This process has made me appreciate light in a way I never could have if it had arrived easily, just as I've grown to appreciate darkness as important in ways that light pollution attempts to hide. When I share this with Luke, he nods. "That cello player didn't seem very impressed when they made an ember with their first go. They were like, 'Okay, what's next?'"

Luke gives me a solemn look. "How do you feel in your body now?" he asks.

I'm sitting beside flames, yet I feel internally cooler. There is no stiletto-heel point of blue light pressing against my brain. I no longer feel like a container full of static electricity.

The rain never really came. The sun has since gone away. In

my body, in the world around me, dark and light have gently been parsed out. They both have a place. And I am sitting in the space between.

My fingers burrow into dark ground—like a salamander finding home, like tree roots stabilizing. It's been hours since I took in artificial light, and I've expelled my bodily heat on the shady sides of mountains. I've channeled my human power into the peeling of bark, the friction of sticks.

"I feel like an underground aquifer that's rising through the eye of a spring," I say.

"That sounds nice," Luke replies, even-keeled, as always.

Out of the corner of my eye, I catch sight of a bat seeking roost in Luke's barn. We are approaching Halloween, referred to by some as a thinning of the veil between past and present, living and dead. This is, in Appalachia, still the traditional season of campfires and storytelling. Luke asks if there's anything I'd like to say, in ode to this fire-making occasion. He tells me that people tend to talk about their ancestors after making their first fire. Luke has always found the impulse poignant since so many cultures refer to fire with their respective terms for "grandfather."

The mention of ancestors makes me think of the work of archeologist Holley Moyes, who has, for years, been looking into how the dark zones of caves came to be sacred in religions the world over. The Central American ritual caves she's focused her career on are still used today by people who believe that spirits are more willing to communicate with the living under the protective cover of darkness.

Homo erectus and Neanderthals used fire. Our ancient ancestors have been pushing back darkness since before we were human, possibly for two million years. And, though early humans did not live in dark zones, our species has long turned to the darkest areas of caves for spiritual rites. The cross-cultural tendency is so prevalent, Moyes started wondering if the presence of darkness itself had something to do with the practice.

"Mother, father, uncle, grandfather—they're all roles, categories we create to assign behaviors. They're all in our heads. They're things that, in a culture, are agreed on, and they're an example of what's known as transcendent imaginary thinking," she's said, suggesting that this ability is part of what makes us human, Paleolithic to present.

The more time Moyes spent in caves, the more she started to suspect darkness played a role in how humans perceive and interact with environs. And she started to wonder if, in modern humans, darkness might still shape thinking. So she enlisted the help of cognitive scientists to conduct a small-scale study.

Participants were divided into two rooms: one well lit, the other dark with a small source of illumination. They were asked things like: If, on the anniversary of your grandfather's death, you woke to the smell of his cigar smoke, would you be more inclined to think that it was a coincidence, or would you think it might be a sign letting you know that his spirit was all right?

In the well-lit room, more people chose rational explanations. In the darker room that mimicked fireside environments, people were far more likely to choose supernatural explanations. "Darkness," Moyes says, "has been an active agent in the development of human spirituality, and it plays a role in how we feel, think, and interpret the world." In other words, darkness might help us reach the depths of our humanity.

"Although we avoid darkness—it makes us uncomfortable, it frightens us—we also seek it out, because it frees our minds from the constant barrage of data that demands our attention every minute of every day," she says. "Darkness allows us to re-create our world and think about ourselves in a different way. Darkness allows us to transcend the ordinary and maybe even find the divine within ourselves."

How wild to think that we might glimpse twilight scenes as enchanted because, just as our sense of smell rises at dusk, darkness activates the potency of our own imaginations. Our minds, opening like the petals of jasmine tobacco, dilating like our widening eyes, so that we might be able to inhabit the infinite possibilities of evening. All those seasons ago, among the fireflies of Great Smoky Mountains National Park, when that stranger called out to ask if I was a magical creature, it was likely a sign that, in fading light, she was blooming into her own magical thinking.

Not much is known about how firelight has altered human culture through the ages. But Polly Wiessner, an anthropologist who has spent a great deal of time with modern hunting-and-gathering Ju/'hoansi bushmen in the Kalahari started to wonder, as Moyes did, if there was something about darkness that shapes the way cultures have developed and, to some extent, still function. "For ninety-nine percent of human evolution," she says, "in darkness, by dim firelight, is how our ancestors lived."

Over the course of unrelated research, she'd collected twenty years of conversational recordings in the Kalahari, with notes about which took place in daylight and which took place at night, fireside. Taking inventory, she found that almost everything talked about in daylight fell into a few categories: complaints, criticisms, gossip, and economics. But after dark, nearly all conversations were dominated by storytelling: tales of adventure, local flora, ancestors, and deities that might be called on to help the living. Criticism, complaints, and economic conversations were almost nonexistent. After dark, there was a change of emotional spectrum, a willingness to consider connections to spirit worlds where perceived boundaries disintegrated. It's where singing and dancing and bonding tended to happen, where daytime arguments found reconciliation. It was in darkness that healing rites were enacted and food was shared.

I think most people would agree that campfire talks tend to take on a different tone than conversations in fluorescent-lit rooms. And darkness itself might be an active agent in that variation. As it is, modern humans spend most waking hours under artificial light, stuck in daylight conversations. We are, increasingly, leaving less room for big questions. Maybe because the ones too large for finite answers are the ones that we, as humans, evolved to explore in shadows.

Of course, the bushmen who participated in Wiessner's study are also modern humans, from modern cultures that are continually shape-shifting, as all cultures do. And their fireside ways of being, though still in practice, are now under threat from geographically distant cultures that have been commandeered by daylight thinking almost completely.

In recent years, Wiessner reports that cell phones have begun to surface among Kalahari teenagers, even though satellites have not yet brought reliable data services. There will undoubtedly be logistical benefits when they do. But—even as someone who lives in a place where cell phone coverage is not a given—I suspect that there will also be little-considered costs, starting with the light pollution of satellite constellations overhead; progressing to: who knows? It's only now that humans have started formally looking into how the interplay of darkness and light might alter human thinking.

Already, the process of severing culture from natural night has started to change things in the Kalahari. It's something Wiessner has only noticed recently, included in her research as a footnote. Even though they're not often within cellular range, Kalahari teenagers have already started preferring the concentrated blue light of their phones to the relaxed red spectra of firelight.

Night after night, they walk away from their elders, away from traditional knowledge, away from stories that encode the wisdom of local ecosystems. They wander deserts, alone, cell phone flashlights pointed not at the path ahead, but at their own faces, searching for signals that they never receive. And, just as I see myself in this fire, I recognize myself in them.

Remembering How to Blink

There is an animal behind me. I hear the crunching of leaves, four rustling feet. Luke raises a finger in the air, signaling that he has heard it, too. "Deer, maybe?" I suggest.

"I think it's Fen," Luke says, just as the dog leaps into the ring of light we've created. Fenrir chooses to sit beside me, ears perked. He shifts his weight until he's firmly pressed against my side. I put an arm around him, and, together, we stare into the fire.

"Do you want him to move?" Luke asks.

"No, I love him," I say. And though we've just met, I do. It's my usual reaction to dogs in general. Fenrir seems to like me, too, though he might not have been drawn to me so much as the fire.

Many predators, including undomesticated canines, avoid fire. Humans are the only predators who produce it, which might be why we have, for so long, felt shielded by it. And dogs are wolves who, domesticated by humans, acclimated to warming themselves by the night-light associated with our species. It's extraordinary to think that Fenrir and I have been brought together tonight by the millions of hearth fires and interdependencies that have connected our kind.

Luke named Fenrir after the hulking black wolf of Norse mythology. For a long time, Fenrir, the wolf, is said to have lived in harmony with the gods. He was well on his way to becoming domesticated—that is, a companion rather than an adversary. But as he grew larger, the gods became terrified of his power. He was too large, too unwieldy, so they had an impulse to contain and control him. They began pushing him into submission rather than working in collaboration. And, through their imprisonment and abuse, they manifested the very monstrous force they'd feared. When Fenrir, the wolf, could take no more of their pressure, one evening, at dusk, he reared back and broke loose of the chains holding him in, leading to the mass destruction of his captors.

Fenrir, the dog, gets up to pad around again, settling behind me. His breath steady. The flames in front of us rising and falling.

I mention to Luke that I'd like to keep practicing with a bow drill, and he offers to send me home with the borrowed pieces of his personal kit. But not without a warning. "You know what 'sophomore' means in Latin? 'Wise fool,'" he says. "Practice. But be careful with it."

I'm a human who has gained a little skill, without enough experience for it to have ripened into wisdom that matches my power. This makes me particularly dangerous.

Luke grabs a branch from the pile of fodder we've gathered, and I realize it's something he's been doing regularly. I've been chatting, ruminating, pulling my energy back into my mind, neglecting the nuanced reality of participating in the living world around me.

When I realize he's been doing all the work of feeding the fire, I'm embarrassed. Luke suggests that I shouldn't be. It's a tendency he notices in almost all new fire-making students. Fire-making takes muscle. But tending is one of the hardest things to teach, because it requires sustained awareness of energy exchange. "We've gotten used to enclosed heat and light. We've gotten used to stoves and lightbulbs and microwaves," he says. "They don't grow. They don't shift or need constant attention. They don't breathe. They don't bring to attention the reality that you have to give to receive, but a fire does that naturally."

Our conversation roams freely, until we both start yawning. He offers for me to camp on his land, but I tell him that I need to head home. That's when he asks what I'd like to do with my fire. And it's a question that, strangely enough, I did not see coming.

I've been so focused on producing light that I've given no thought to how I might invite darkness back. I overlooked the fact that, in making fire, I simultaneously created responsibility. This, even though it is darkness that brought me here. This, even though it is darkness that I've come to honor, in a sense.

Subconsciously, I guess I thought I'd just let the flames wilt, smoking themselves to death. Or that I'd douse things with water when it was time to move on. But now that I've come to know firelight with such intimacy, these choices seem neglectful—disrespectful of all this ecosystem has given, dismissive of my own energy.

How extremely human of me to overlook light's ending in favor of its beginning. For the duration of our species' evolution, we've needed to know how to build and sustain fires to survive life on this planet. Even now, creating and controlling fire is, in some capacity, the basis of our most advanced technologies.

We look to light to save us, because, until now, it almost always has. Evolution has coded us to be attracted to fire, to cultivate it. And now that it's time to turn down artificial light and the heat of all our internal combusting, we're collectively struggling to figure out how. Because moving toward darkness, rather than light, requires an about-face reversal of the human story. It requires flip-the-script storytelling. And, at this critical juncture in history, natural night needs human help to regain its rightful place on Earth, as well as in the heavens.

I cannot save the planet from light pollution. But I'm starting to understand what's going on in my own life, within myself. There is, in biology, a "super stimulant" concept, which explains that, when an animal has a natural tendency to react to a stimulant, an exaggerated form of that stimulant will cause the animal to exaggerate its reaction to match. It has been studied among birds and was popularized when fish, evolutionarily coded to react to the red underbelly of another species, were found to go exceptionally wild over giant pieces of painted red wood. Human gravitation toward junk food is often used as an example since intake of high-energy foods would have once given evolutionary advantage, so we crave as much as we can get. But I'm not sure I've ever experienced anything as super-stimulant strong as artificial light—it calls to me in its various forms, and gorging on it from the comfort of my couch is as easy as chomping chips.

When moths circle lightbulbs, throwing themselves against sources of artificial light, they are likely aiming for the moon. When birds get caught in tunnels of light, they're likely trying to read the stars. And when we throw ourselves against computers and tablets and phones as we scroll social media sites, we are likely attempting to reach the hearth fires that we, as humans, have always depended on. We're seeking sensory-rich exchanges and comradery and caring reconciliation. And, though our minds might be stimulated by their abstract representations, our bodies are finding two-dimensional, amped-up facades. Artificial light, a false prophet.

It seems so simple: Let's power down our computers! Let's put away our phones! Let's stop taking in passive entertainment and make our own! Let's turn the lights off so we can get to sleep! And yet, many of us can't.

Reducing light pollution—on both micro and macro levels—doesn't fundamentally require human technology, only human temperance. After all, we have the technology we need to solve many of our most pressing environmental issues, across disciplines, but we have a hard time culturally accepting and implementing them. We argue about economics. We tussle about resources. We complain and criticize. We get stuck in our daylight minds when what we need are reduced-energy conversations of aid and healing. In lieu of hearth fires, we turn to online forums, staring directly into the blue light that our bodies have potentially evolved to associate with logistics and griping, light spectra that studies have shown to increase emotions of dissatisfaction and depression.

Lightbulbs are ubiquitous symbols of epiphany. But what if darkness, not light, is the mother of meaning? There seems to be a growing awareness that, with the loss of natural night, we're missing something more than opportunities to stargaze. There has been a rise in the popularity of "dark dining," where people eat without lights so that they might tap their nonvisual senses. There are, increasingly, darkness retreats, where people go into light-sealed rooms for weeks at a time, food delivered through door slits like alcohol in a Prohibition speakeasy. Unable to sip darkness as nature dictates, on those retreats, people attempt to chug it as if to make up for the deficiency. But we cannot be sustained by darkness or light alone. It is only through the impermanence of day and night that life on Earth endures.

Maybe humanity isn't afraid of the dark so much as we're afraid of losing light. It's been our species' caretaker for so long, it's understandable that we don't want to let it out of our sight. But when you know how to make light with your own hands, when you know that you are the light, it's easier to welcome night. It's easier to accept that a dark screen is not the end of a story. It's more of a meditative passage in an ongoing narrative in which we are, as humans, key players. Because we all—every single one of us—have fire in our bellies. We all, in various combustive forms, have flames at our fingertips.

The curious thing about this hearth fire—as opposed to all the electric lights in my life—is that, if it went out completely, I could take the outstretched hand of a spicebush shrub or tulip poplar tree and they'd lead me back to light. It's wild to befriend a living forest in this way. And after experiencing the elation of fire-making, after living out the true energetic cost of having night light, it's hard to consider this fire's demise. But, just as there is an art to building a blaze, there is an art to dismantling it with grace.

Luke tells me that the most elegant way he knows to put a fire out is also an exercise in human restraint. It's more ceremony than convenience. It's not a single action, but, rather, the accumulation of many. In terms of animal behavior, the fact that humans can extinguish fires might be no less amazing than the fact that we can build them. Because, sometimes, to survive, we need to build fires. And, sometimes, to survive, we need to put fires out.

To travel toward the deep end of darkness with paleo-elegance, Luke explains that I need to invest time in retracing my steps, moving through the tinder spectrum in the opposite direction, enacting a light weaning. I need to nurse the flames slowly, adding ever-smaller branches, then twigs, until my offerings become so tiny that I can barely feel them between my fingertips. When I focus my bodily energy on tending this fire down, the flames I've built will naturally settle back into embers.

By the lowering light of this fire, I'm inspired. In the coming weeks, I will attempt to purchase some incandescent, warm-spectrum bulbs for my bedroom lamps, only to discover that they've already been phased out of the market—due to concerns about the heat of energy, without regard to the rising cost of light. By necessity, then, I'll go all in, buying bulbs that burn campfire red. The change of color itself will inspire me to notice the presence of indoor lighting I've previously taken for granted as no more noticeable than air.

The bulb's red-hot hue will reconcile, in my mind if not my

body, the realities of heat and light. It will remind me that, for light's presence at night, I am always trading natural darkness. I am always destroying something; that's the cost of artificial light. I'm not interested in trying to hand-make fire every evening, but becoming more cognizant of energy will lead me to, in my nightly habits, wean from blue spectra a little earlier, switching to warm lights that match sunset-ripening skies in real time.

This will help me get out of my circadian rhythm's way, let my body freely produce melatonin and rhodopsin, give myself space for far-roaming ideas as I drift. I'll soon start sleeping better. Whether it's a physiological response or a psychological reaction, I cannot say—and I'm not sure it matters. Either way, I'll be sliding into night on sunset beams, reading books by light that evokes campfires.

Archer, in his preteen way, will complain that the red bulbs make our house look haunted. I'll welcome the teachable moment, pointing out that every lightbulb powered by fossil fuel is a portal through which ghost energy enters. A lightbulb powered by coal is the resurrected energy of plants that brushed against the legs of dinosaurs. A lightbulb fed by oil marks the energetic return of creatures that once floated in ancient inland seas. Even solar energy is the stored memory of a sun that's left the scene, if temporarily.

The less light used, the lesser the haunting. I imagine how I might have explained this to him as a small child, how it might have helped him see artificial light as more of an overarching threat than natural darkness from the start.

I'll think he's written the whole thing off until I see red light coming from under his bedroom door. Then, one evening, a question: "What's better for sleep, a red light on or no light at all?" And, that very night, he'll change his pattern of keeping a night-light on, choosing, instead, to sleep in the dark.

This gentler arrival of night indoors will help me see my bedroom anew. Just as I've come to know my neighborhood differently at night, I'll start noticing tiny blue and green lights on my device chargers, which make my bedroom look less like a sacred cave, more like a spaceship control panel. Studies have shown that even low-level artificial light can disturb sleep patterns and activate the body's fight-or-flight responses instead of allowing for parasympathetic, cooling states that create slowed heartbeats and relaxed breathing during sleep.

In Japan, researchers have found that the brightness of a single streetlight shining through a bedroom window increases rates of depression. In China, they have found that light pollution is linked to diabetes. Even sleeping with the light of a television screen is enough to cause glucose imbalances that contribute to weight gain, obesity, and cardiometabolic disease. Our bodies gauge light and dark even when we're unconscious, tremendously altering bodily processes in ways we're just discovering. For these reasons and more, I'll unplug most of the devices in my bedroom. The few light-dots that seem necessary to leave plugged in, I'll cover with pieces of painter's tape.

These actions will not clear all the modern-day tinder from my mind. They will not prevent all my fritzed-out days. They will not save me from feeling behind. They will not stop me from checking email or falling prey to Netflix binges. They will not stop the world from spinning. If anything, they will help my body better sense rotation: darkness, slipping into residence so that daylight can take a break.

In time, both indoor and outdoor light shifts will be felt in my body, with regularity, as a minor-key settling. I will, finally, be showing myself the same courtesy that I exhibited toward fireflies when I used red lights in their domain. It has taken all of this to help me see how, when I pulled curtains tight to protect fireflies, I was wrapping all that artificial light around myself. I knew then that the spectra that are best for fireflies at night— that is, red lights or no lights at all—were also best for me. Still, it took making fire with sticks and my bare hands for me to absorb the fact that I have the capacity to release my grip on high-octane light energy.

Exploring the dark side of these mountains—on a journey to find beauty in the yin as well as the yang—has ultimately prevented me from living a life of lights unexamined. Maybe I was always meant to end up here, a student of darkness studying light in the form of a dwindling fire that's still warming my hands.

~~~~~~~~~~

When a neighbor shows up for a visit, his arrival is announced via LED headlights in the distance. The lights serve as a reminder that we are visiting, not living in, the Paleolithic. Luke leaves to greet the new arrival. Fenrir and I will follow, but we linger for a minute in prehistory, enjoying our shared ancestry. Me, of the fire-making hominids. Him, of the fire-sitting wolves who roamed the late Pleistocene.

This happens to be the height of the autumnal songbird migration, bookend to last spring's owl-moonwatching. I imagine that from above, this fire looks like a tiny red ember at the base of a massive, charred-spindle-dust mountain. If I listen carefully, I might catch wind of a thousand birds, headed to a thousand different nesting places, all of them protected by this relatively dark Appalachian flyway.

Every marvel I've ever searched for in these mountains already knew where to find me. There are surely leopards and tigers among this acreage's holy humus. And I have good reason to believe that, along a nearby creek, dormant glowworms are waiting for the return of their luminescent season. And on every uncertain path I've wandered, in every moment of bewilderment and wonder, even on the darkest of nights when I've felt lost in grief and confusion, I've been carrying the promise of light in my core, sure as a firefly illuminates the scale of their world.

Those creatures, like me, are only able to make light because they are blessed by the offerings of other species that, like them, depend on cycles of light and dark—right down to the sturdy oak leaves that give them shelter. Firefly larvae are undoubtedly encircling this hearth, their lanterns shining and fading. All of us preparing for the long night of winter. Unlike in years past—now that I have leaves and stars tangled in my hair—I do not dread it.

From the cycles of light and dark, none of us can be parsed out completely. On and on, we all, together, keep bringing this world into being. Night follows day. Spring follows winter. How fortunate I am to be among the humans who've witnessed this shape-shifting, this perpetual blinking. I hope I am not among the last. We craft the world, bulb by bulb, seed by seed. And we'll know we're on our way to wellness when stars begin to, once again, reveal themselves.

May we learn to love darkness as our ancestors learned to love light, so that we might play a role in nature's reliable cycling. May we begin to recognize that, just as we've tended the lights up, we can tend them down—revealing wonders that are, in daylight, unimaginable. May we find our way back to natural darkness, or at least hold fast to the wilderness that still exists, so that we'll be able to bear witness to night's living riches. May we, as a species, relearn how to blink, letting both night and day have their space. Because it is only by the power of light and the grace of darkness that we're able to rest and rise, then rest and rise again. That's the beauty; that's the blinking.

Behind me, Fenrir stirs. In front of me, newly formed bits of charcoal crackle and cackle, energy condensing in their bright-hot cores. I'm tending this fire down to darkness, but I know that if I called on flames, they would rush back to my side, sure as Fenrir. At this point, all it would take is a whisper. There is comfort in the notion. There is power in resisting the urge. I lean into the black dog at my back, and he presses his weight against mine. We are animals, both acutely alive and resting in peace tonight.

Acknowledgments

~~~~~~

I am indebted to fireflies. When I wrote an article about a synchronous-blinking species for the *Washington Post Magazine*, many readers reached out to let me know that they'd started turning off their porch lights more often. They were curious to discover what they'd been missing out on by not inviting darkness into their own yards, and I was amazed that my story had inspired real-world action that led to reduced light pollution. I appreciate everyone who let me know that story meant something to them. Thanks also to Richard Just and my longtime editor, David Rowell, who has entertained my wild story ideas for nearly twenty years.

Heather Carr has been a steadying force during this long journey through the dark. I'm deeply grateful for her guidance, as well as the support of Molly Friedrich, Lucy Carson, Hannah Brattesani, and Marin Takikawa. This book has received above-and-beyond care from Amy Gash—a thoughtful editor and kind soul who understood this project from the beginning—as well as Betsy Gleick, Debra Linn, Marisol Salaman, Christopher

Moisan, Steve Godwin, Martha Cipolla, Brenna Franzitta, Laura Essex, Tom Mis, and others working behind the scenes.

I'm blessed to have Randall and Carolyn Henion as parents. They let me free-roam creeks and forests as a child—and then watched with minimal judgment as I ambled nontraditional career paths as an adult. They have been incomparable guides in every aspect of life. Special thanks to Archer for his willingness to (sometimes) roam mucky places with me. Archer, no matter where you go or how urbane your interests get, I hope you'll always embrace these living mountains as home.

In the process of working on this book, I was often so wowed by what I found that I had the impulse to immediately reach out to friends because I couldn't wait to share. There were tidbits in every chapter that made me think, *What? I've got to tell someone about this!* If you ever geeked out with me over salamander eggs or moth behavior, I remember and appreciate your enthusiasm. A special nod to the Lippards, Jaremas, Polings, Marshes, Russells, Peters, Bookwalters, McAllisters, and Penningtons. Also to Matt and the entire Hrenak family.

Lori Williams offered helpful advice when my fieldwork required pivoting, and—in addition to the many gracious individuals who appear in this book—I appreciate the assistance of Caitlin Worth, Dana Soehn, Allison Cochran, Landis Taylor, and Jesse Pope. Science is always evolving, but I've worked to synthesize complex information as it is currently understood. Thanks to Christopher Kyba for leading me in the right direction and to John Barentine, Gary Walker, Travis Longcore, and Avalon Owens for providing feedback on early drafts.

I owe a great deal to the editors I have worked with throughout my career, as well as academic colleagues including Joseph

Bathanti, Sandy Ballard, Susan Weinberg, Mark Powell, and Betty Conway. I'm grateful to David Joy, Alyssa Tsagong, Sunny Townes, Bethany Jewell Gray, Amy Cooke, Mike Reynolds, and Denise Powell for cheering me on when I needed it. And I applaud my capstone creative writing students for developing a sense of community and revelry that once inspired a passerby to ask if we'd relocated from the drama department.

My time as an Alicia Patterson fellow was formative in my development as a thinker and writer, and I'm grateful to Margaret Engel for seeing me through the challenges of 2020. *Night Magic* has been supported by the Alfred P. Sloan Foundation, and I appreciate the work of Doron Weber, Shriya Bhindwale, and Peter DiFranco in facilitating that process.

Thanks to Wendell Berry for allowing me to share his words as an epigraph—and for reminding me that there is great pleasure in corresponding via post. It had been a long time since I'd reached into the dark hollow of a mailbox to find a handwritten message. What a gift to recognize that, when we choose to, we can still conduct business at the speed of poetry—that is to say, the speed of stamps and Wendell Berry.

It's impossible to know where *Night Magic* will go from here. The one thing I know for sure is that releasing this book is not an end; it is a beginning. Thank you, dear reader, for being part of *Night Magic*'s ongoing story.

Selected Bibliography

Fireflies Blinking

Chávez, Karen. "Discovery of synchronous fireflies at Grandfather Mountain could help ease Smokies crowds." *Asheville Citizen-Times*, 3 September 2019. citizen-times.com/story/life/2019/09/03/synchronous-fireflies-have-been-discovered-grandfather-mountain-nc/2165336001/

Chepesiuk, Ron. "Missing the Dark: Health Effects of Light Pollution." *Environmental Health Perspectives*, January 2009, pp. A20–A27. doi/10.1289/ehp.117-a20

Doll, Jen. "The Great Smoky Mountains' Incredible Firefly Light Show." *Mental Floss*, 9 June 2014. mentalfloss.com/article/57188/great-smoky-mountains-incredible-firefly-light-show

Dunn, Nick, and Tim Edensor. *Rethinking Darkness: Cultures, Histories, Practices.* Routledge, 2021.

Falchi, Fabio, et al. "The new world atlas of artificial night sky brightness." *Science Advances*, vol 2, no. 6, 2016. Doi:10.1126/sciadv.1600377

Faust, Lynn. *Fireflies, Glowworms, and Lightning Bugs: Identification and Natural History of the Fireflies of the Eastern Central United States and Canada.* University of Georgia Press, 2017.

Lewis, Sara M., et al. "Firefly tourism: Advancing a global phenomenon toward a brighter future." *Conservation Science and Practice*, vol. 3, no. 5, 2021, p. e391. doi/full/10.1111/csp2.391

Lewis, Sara. *Silent Sparks: The Wondrous World of Fireflies*. Princeton University Press, 2016.

"Synchronous Fireflies." *Learn About the Park*. National Park Service. nps.gov/grsm/learn/nature/fireflies.htm

Trimmer, B. A., et al. "Nitric oxide and the control of firefly flashing." *Science*, June 2001. doi10.1126/science.1059833

Salamanders Migrating

Baars, Samantha. "Walk like an amphibian: The spotted salamander gets a little help from friends." *C-Ville Weekly*, 15 February 2017. c-ville.com/walk-like-amphibian-spotted-salamander-gets-little-help-friends/

Bedrosian, T. A., and R. J. Nelson. "Timing of light exposure affects mood and brain circuits." *Translational Psychiatry*, 7 January 2017, p. e1017. doi/10.1038/tp.2016.262

Berenbaum, May. "Sea Monkey® See, Sea Monkey® Do." *American Entomologist*, vol. 45, no. 2, 1 April 1999, pp. 68–69. doi.org/10.1093/ae/45.2.68

Binkovitz, Leah. "The Surprisingly Colorful Salamanders of Appalachia." *Smithsonian Magazine*, 18 June 2013. smithsonianmag.com/smithsonian-institution/the-surprisingly-colorful-salamanders-of-appalachia-813148/

Burns, Angus C., et al. "Day and night light exposure are associated with psychiatric disorders: an objective light study in >85,000 people." *Sleep*, vol. 46, no. 1. 29, May 2023, p. A136. doi.org/10.1093/sleep/zsad077.0307

Castañón, Laura. "This salamander can regenerate limbs like Deadpool. Can it teach us to do the same?" *Northeastern Global News*. 22 July 2019. northeastern.edu/2019/07/22/heres-what-we-can-learn-from-a-salamander-that-can-regenerate-its-limbs/

Edwards, Nina. *Darkness: A Cultural History*. Reaktion Books, 2018.

For, Jonathan. "Salamander Regeneration as a Model for Developing Novel Regenerative and Anticancer Therapies." *Journal of Cancer*, 2014, pp. 715–719. doi/10.7150/jc.9971

Jones, Benji. "The animal that's everywhere and nowhere." *Vox*, 25 January 2022. vox.com/22877353/axolotl-salamander-pet-extinction-mexico

Kauer, John. "On the scents of smell in the salamander." *Nature*, vol. 417, 16 May 2002, pp. 336–342. doi.org/10.1038/417336a

Kerney, Ryan, et al. "Intracellular invasion of green algae in a salamander host." *Proceedings of the National Academy of Sciences*, April 2011, p. 108. doi/10.1073/pnas.1018259108

Li, Zhenlong, et al. "Blue light at night produces stress-evoked heightened aggression by enhancing brain-derived neurotropic factor in the basolateral amygdala." *Neurobiology of Stress*, vol. 28, January 2024. doi/10.1016/j.ynstr.2023.100600

Merchant, Brian. "The Man Who Turned Night Into Day." *Vice*, 20 January 2016. vice.com/en/article/9a3y8d/the-man-who-turned-night-into-day

Morrow, Erica, et al. "The Chicago Alley Lighting Project: Final Evaluation Report." *Illinois Criminal Justice Information Authority*, April 2000. ojp.gov/ncjrs/virtual-library/abstracts/chicago-alley-lighting-project-final-evaluation-report

Paksarian, Diana, et al. "Association of Outdoor Artificial Light at Night With Mental Disorders and Sleep Patterns Among US Adolescents." *JAMA Psychiatry*, 2020, p. 77. doi/10.1001/jamapsychiatry.2020.1935

Pearson, Gwen. "Here Be Dragons." *Wired*, 18 December 2013. wired.com/2013/12/the-secret-underwater-world-of-dragons/

Rich, Catherine, and Travis Longcore, editors. *Ecological Consequences of Artificial Night Lighting*. Island Press, 2006.

Steinbach, Rebecca, et al. "The effect of reduced streetlighting on road casualties and crime in England and Wales: controlled interrupted time series analysis." *Journal of Epidemiology & Community Health*, 2015, pp. 1118–1124. jech.bmj.com/content/69/11/1118

Stolzenberg, Lisa, et al. "A Hunter's Moon: The Effect of Moon Illumination on Outdoor Crime." *American Journal of Criminal Justice*, vol. 42, 16 June 2016, pp. 188–197. doi.org/10.1007/s12103-016-9351-9

Van Elk, Michiel, et al. "The neural correlates of the awe experience: Reduced default mode network activity during feelings of awe." *Human Brain Mapping*, vol. 40, no. 12, 15 August 2019, pp. 3561–3574. doi/full/10.1002/hbm.24616

Welsh Jr., Hartwell H., and Garth R. Hodgson. "Woodland salamanders as metrics of forest ecosystem recovery: a case study from California's redwoods." *Ecosphere*, vol. 4, no. 5, 2013, pp. 1–25. doi/full/10.1890/ES12-00400.1

Wise, Sharon, et al. "The Effects of Artificial Night Lighting on Tail Regeneration and Prey Consumption in a Nocturnal Salamander (*Plethodon cinereus*) and on the Behavior of Fruit Fly Prey (*Drosophila virilis*)." *Animals*, 17 August 2022. doi/10.3390/ani12162105

Owls Nesting

"Ancients: Owls Through the Years—A Look at the Evolution of Athenian Tetradrachms." Numismatic Guaranty, 7 December 2011. ngccoin.com/news/article/2245/Ancients-Owls-Through-the-Years---A-Look-at-the-Evolution-of-Athenian-Tetradrachms/

"Appliance Standards Rulemakings and Notices." U.S. Department of Energy. www.1.eere.energy.gov/buildings/appliance_standards/standards.aspx?productid=4

Baker, M. C. "Bird Song Research: The Past 100 Years." *Bird Behavior*, vol. 14, January 2001. researchgate.net/publication/263107344_Bird_Song_Research_The_Past_100_Years

"Eastern Screech-Owl Life History." *All About Birds*. Cornell Lab of Ornithology. www.allaboutbirds.org/guide/Eastern_Screech-Owl/lifehistory

Gracheva, Elena O., et al. "Molecular basis of infrared detection by snakes." *Nature*, vol. 464, 14 March 2010, pp. 1006–1011. nature.com/articles/nature08943

Heisman, Rebecca. "Owl Be Seeing You: Amazing Facts About Owl Eyes." *Bird Calls*, American Bird Conservancy. abcbirds.org/blog/owl-eyes

Hölker, Franz, et al. "The Dark Side of Light: A Transdisciplinary Research Agenda for Light Pollution Policy." *Ecology and Society*, vol. 15, no. 4, December 2010. jstor.org/stable/26268230?seq=3

"In a Different Light." NASA. www.science.nasa.gov/mission/hubble/science/science-behind-the-discoveries/wavelengths/

Jimenez, Mikko. "Where Will Rocky the Northern Saw-whet Owl Spend the Holidays?" *Migratory Bird Initiative*, National Audubon Society, 22 December 2020. audubon.org/news/where-will-rocky-northern-saw-whet-owl-spend-holidays

Job, Jacob. "They Tap Into the Magical, Hidden Pulse of the Planet, but What is the Nighttime Bird Surveillance Network?" *Scientific American*, 18 August 2023. scientificamerican.com/podcast/episode/they-tap-into-the-magical-hidden-pulse-of-the-planet-but-what-is-the-nighttime-bird-surveillance-network/

Longcore, Travis, et al. "Rapid assessment of lamp spectrum to quantify ecological effects of light at night." *Journal of Experimental Zoology Part A: Ecological and Integrative Physiology*, vol. 329, no. 8–9, 12 June 2018, pp. 511–521. doi.org/10.1002/jez.2184

Love, Sarah J. "Sky Islands Are a Global Tool for Predicting the Ecological and Evolutionary Consequences of Climate Change." *Annual Review of Ecology, Evolution, and Systematics*, vol. 54, 4 August 2023, pp. 219–236. 10.1146/annurev-ecolsys-102221-050029

Patel, Kasha, et al. "LED lights are meant to save energy. They're creating glaring problems." *Washington Post*, 23 June 2023. washingtonpost.com/climate-environment/interactive/2023/glaring-problem-how-led-lights-worsen-light-pollution/

Romps, David M., et al. "Projected increase in lightning strikes in the United States due to global warming." *Science*, 14 November 2014, vol. 346, no. 6211, pp. 851–854. doi/10.1126/science.1259100

"Safe and Effective Wildfire Response." U.S. Department of Agriculture. www.fs.usda.gov/managing-land/fire/response

Sánchez de Miguel, Alejandro, et al. "Environmental risks from artificial nighttime lighting widespread and increasing across Europe." *Science Advances*, vol. 8, no. 37, 14 September 2022. 10.1126/sciadv.abl6891

Sánchez de Miguel, Alejandro, et al. "First Estimation of Global Trends in Nocturnal Power Emissions Reveals Acceleration of Light Pollution." *Remote Sensing*, vol. 13, no. 16, 21 August 2021. doi/10.3390/rs13163311

Scott, W. E. D. "Some Observations on the Migration of Birds." *Nature*, vol. 24, 21 July 1881. nature.com/articles/024274a0

Senzaki, Masayuki, et al. "Sensory pollutants alter bird phenology and fitness across a continent." *Nature*, vol. 587, 11 November 2020, pp. 605–609. nature.com/articles/s41586-020-2903-7

Sheppard, Christine, and Bryan Lenz. "Birds Flying Into Windows? Truths About Birds & Glass Collisions from ABC Experts." American Bird Conservancy, 2023. www.abcbirds.org/blog/truth-about-birds-and-glass-collisions/

Sordello, Romain, et al. "A plea for a worldwide development of dark infrastructure for biodiversity—Practical examples and ways to go forward." *Landscape and Urban Planning*, vol. 219, March 2022. doi.org/10.1016/j.landurbplan.2021.104332

"Spectroscopy: Reading the Rainbow." NASA. www.hubblesite.org/contents/articles/spectroscopy-reading-the-rainbow

Stark, H., et al. "Nighttime photochemistry: nitrate radical destruction by anthropogenic light sources." American Geophysical Union. Fall Meeting 2010. ui.adsabs.harvard.edu/abs/2010AGUFM.A21C0117S/abstract

Stelloh, Tim. "Tiny owl rescued from Rockefeller Center Christmas tree that traveled 170 miles to NYC." NBC News, 18 November 2020. nbcnews.com/news/animal-news/tiny-owl-rescued-rockefeller-center-christmas-tree-travelled-170-miles-n1248166

Villing, Alexandra, and Susan Deacy. *Athena in the Classical World*. Brill, 2001.

"What is the visible light spectrum?" NASA. www.science.nasa.gov/ems/09_visiblelight/

"Why Is the Sky Blue?" *SciJinks*. National Oceanic and Atmospheric Administration. www.scijinks.gov/blue-sky/

Glowworms Squirming

De Guzman, Edith. "2030 Landscapes: Shade in LA—Rising Heat Inequity in a Sunburnt City." Sierra Club, 8 June 2022. sierraclub.org/articles/2022/06/2030-landscapes-shade-la-rising-heat-inequity-sunburnt-city

DeVille, Nicole V., et al. "Time Spent in Nature is Associated with Increased Pro-environmental Attitudes and Behaviors." *International Journal of Environmental Research and Public Health*, 18 July 2021, p. 7498. doi/10.3390/ijerph18147498

Falaschi, Rafaela. "*Neoceroplatus betaryiensis* nov. sp. (Diptera: Keroplatidae) is the first record of a bioluminescent fungus-gnat in South America." *Nature Scientific Reports*, 5 August 2019, p. 11291. nature.com/articles/s41598-019-47753-w

Fulton, B. B. "A Luminous Fly Larva with Spider Traits (Diptera, Mycetopilidae)." *Annals of the Entomological Society of America*, vol. 34, no. 2, 1 June 1941, pp. 289–302. doi/10.1093/aesa/34.2.289

Gaynor, Kaitlyn M., et al. "The influence of human disturbance on wildlife nocturnality." *Science*, vol. 360, no. 6394, 15 June 2018, pp. 1232–1235. doi/10.1126/science.aar7121

Henry, Cait M., et al. "It Felt Like Walking Through a Night Sky: Managing the Visitor Experience During Biologically-based Nighttime Events." *Event Management*, March 2022, p. 387–403. doi/3727/152599 521X16288665119314

Lemelin, Raynald Harvey, et al. "Entomotourism: The Allure of the Arthropod." *Society and Animals*, October 2019, pp. 733–750. doi/10.1163/15685306-00001830

Lemelin, Raynald Harvey, et al. "The Bioluminescent Insects of Grandfather Mountain Stewardship Foundation." *Tourism Cases*, June 2023. doi.org/10.1079/tourism.2023.0014

McDonald, Robert I., et al. "The tree cover and temperature disparity in US urbanized areas: Quantifying the association with income across 5,723 communities." *PLOS ONE*, 28 April 2021. doi.org/10.1371/journal.pone.0249715

Rockhill, Aimee P., et al. "The Effect of Illumination and Time of Day on Movements of Bobcats (*Lynx rufus*)." *PLOS ONE*, July 2013, vol. 8, no. 7, p. e69213. doi/10.1371/journal.pone.0069213

Sorenson, Clyde. "Hunting Carolina Ghosts." *Our State*, 27 March 2023. ourstate.com/hunting-carolina-ghosts/

Yin, Yi, et al. "Unequal exposure to heatwaves in Los Angeles: Impact of uneven green spaces." *Science Advances*, 28 April 2023, vol. 9, no. 17. doi/10.1126/sciadv.ade8501

Viviani, Vadim R., et al. "A new brilliantly blue-emitting luciferin-luciferase system from *Orfelia fultoni* and Keroplatinae (Diptera)." *Nature Scientific Reports*, 15 June 2020, p. 9608.

Moths Transforming

Alcock, John, and Gary Dodson. "The Diverse Mating Systems of Hilltopping Insects." *American Entomologist*, Summer 2008, pp. 80–87. doi/10.1093/ae/54.2.80

Bittel, Jason. "Nocturnal pollinators go dark under street lamps." *Nature*, 2 August 2017. nature.com/articles/nature.2017.22395

Boeckmann, Catherine. "Do Woolly Worms Really Predict Winter Weather?" *Old Farmer's Almanac*, 27 November 2023. almanac.com/woolly-bear-caterpillars-and-weather-prediction

Bogard, Paul. *The End of Night: Searching for Natural Darkness in an Age of Artificial Light*. Little, Brown and Company, 2013.

Borough and Precinct Crime Statistics. New York City Police Department. nyc.gov/site/nypd/stats/crime-statistics/borough-and-precinct-crime-stats.page

"Countries Urged to Prioritize Protection of Pollinators to Ensure Food Security at UN Biodiversity Conference." The United Nations, 6 December 2016. un.org/sustainabledevelopment/blog/2016/12/pollinators/

Deitsch, John F., and Sara A. Kaiser. "Artificial light at night increases top-down pressure on caterpillars: experimental evidence from a light-naïve forest." *Proceedings of the Royal Society*, 8 March 2023. doi/10.1098/rspb.2023.0153

Ezzell, Tim, et al. "Case Study: Point Pleasant, West Virginia." *Trends and Strategies for Tourism in Appalachia*, Appalachian Regional Commission, March 2020, pp. 70–75. arc.gov/wp-content/uploads/2021/02/ARC-Tourism-report-final-Dec-2020-1.pdf

Goulson, Dave. "The insect apocalypse: Our world will grind to a halt without them." *Guardian*, 25 July 2021. www.theguardian.com/environment/2021/jul/25/the-insect-apocalypse-our-world-will-grind-to-a-halt-without-them

"Leaf Miners: the Inside Story." *Garden Stories*. Chicago Botanic Garden. chicagobotanic.org/blog/plant_science_conservation/leaf_miners_inside_story

Li, Yanwei, et al. "A native sericin wound dressing spun directly from silkworms enhances wound healing." *Colloids and Surfaces B: Biointerfaces*, vol. 225, May 2023, p. 13228. doi/10.1016/j.colsurfb. 2023.113228

Macgregor, Callum J., et al. "The dark side of street lighting: impacts on moths and evidence for the disruption of nocturnal pollen transport." *Global Change Biology*, 1 June 2016. doi/10.1111/gcb.13371

McCormac, Jim. "The hickory horned devil is one giant caterpillar." *Columbus Dispatch*, 16 September 2022. dispatch.com/story/lifestyle/home-garden/2022/09/16/caterpillars-are-a-fascinating-part-of-moths-four-stage-life-cycle/68188212007/

McCormac, Jim, and Chelsea Gottfried. *Gardening for Moths: A Regional Guide*. Ohio University Press, 2023.

Robertson, Stephen M. "Nocturnal Pollinators Significantly Contribute to Apple Production." *Journal of Economic Entomology*, vol. 114, no. 5, October 2021, pp. 2155–2161. doi/10.1093/jee/toab145

"Species Spotlight Cecropia Moth." National Park Service. nps.gov/articles/species-spotlight-cecropia-moth.htm

Stevenson, Alexa. "Why are moths attracted to light?" *Penn State News*, 19 October 2008. psu.edu/news/research/story/probing-question-why-are-moths-attracted-light/

Tanizaki, Jun'ichirō. *In Praise of Shadows*. 1933. Leete's Island Books, 1977.

Travouillon, Kenny J., at al. "All-a-glow: spectral characteristics confirm widespread fluorescence for mammals." *Royal Society Open Science*, 4 October 2023. doi/10.1098/rsos.230325

Van Doren, Benjamin M., et al. "Drivers of fatal bird collisions in an urban center." *Proceedings of the National Academy of Sciences*, 7 June 2021, p. 118. doi/full/10.1073/pnas.2101666118

Van Doren, Benjamin M., et al. "High-intensity urban light installation dramatically alters nocturnal bird migration." *Proceedings of the National Academy of Sciences*, 2 October 2017, p. 114. doi/10.1073/pnas.1708574114

Van Geffen, Koert G., et al. "Artificial light at night causes diapause inhibition and sex-specific life history changes in a moth." *Ecology and Evolution*, 25 April 2014. doi/10.1002/ece3.1090

Van Langevelde, Frank. "Artificial night lighting inhibits feeding in moths." *Biology Letters*, 1 March 2017. doi/10.1098/rsbl.2016.0874

Winn, Marie. *Central Park in the Dark: More Mysteries of Urban Wildlife*. Farrar, Straus and Giroux, 2008.

Bats Flying

Adams, Joshua, et al. "Success of Brandenbark™, an artificial roost Structure Designed for use by Indiana Bats." *Journal of the American Society of Mining and Reclamation*, vol. 4. doi/10.21000/jasmr15010001

Calma, Justine. "Giving a bat flowers might preempt a pandemic." The Verge, 16 November 2022. https://www.theverge.com/2022/11/16/23461080

Celly, Courtney. "Bats are one of the most important misunderstood animals." U.S. Fish and Wildlife Service. www.fws.gov/story/bats-are-one-most-important-misunderstood-animals

Chen, Caroline. "The Scientist and the Bats." *ProPublica*, 22 May 2023. www.propublica.org/article/australia-bats-hendra-research-pandemic-prevention

Cheng, Tina L., et al. "The scope and severity of white-nose syndrome on hibernating bats in North America." *Conservation Biology*, 20 April 2021. doi.org/10.1111/cobi.13739

Eby, Peggy, et al. "Pathogen spillover driven by rapid changes in bat ecology." *Nature*, vol. 613, 16 November 2022, pp. 340–344. doi.org/10.1038/s41586-022-05506-2

Lu, Manman, et al. "Does public fear that bats spread COVID-19 jeopardize bat conservation?" *Biological Conservation*, vol. 254, February 2021. doi.org/10.1016/j.biocon.2021.108952

O'Shea, Thomas J., et al. "Bat Flight and Zoonotic Viruses." *Emerging Infections Diseases*, vol. 20, no. 5, May 2014, pp. 741–745. doi.org/10.3201/eid2005

Palmer, Brian. "The case for spider conservation: They keep pests from devouring humans' food supply." *Washington Post*, 21 July 2014. www.washingtonpost.com/national/health-science/the-case-for-spider-conservation-they-keep-pests-from-devouring-humans-food-supply

Pinson, Jerald. "What is it good for? Absolutely one thing. Luna moths use their tails solely for bat evasion." *Research News*. Florida Museum of Natural History. www.floridamuseum.ufl.edu/science/what-is-it-good-for-absolutely-one-thing-luna-moths-use-their-tails-solely-for-bat-evasion/

"Rabies." *One Health*, American Veterinary Medical Association. avma.org/resources-tools/one-health/rabies

Shen, Zhiyuan, et al. "Biomechanics of a moth scale at ultrasonic frequencies." *Biophysics and Computational Biology*, 12 November 2018, pp. 12200–12205. doi/10.1073/pnas.1810025115

Straka, Tanja M., et al. "Tree Cover Mediates the Effect of Artificial Light on Urban Bats." *Frontiers in Ecology and Evolution*, vol. 7, 2019. doi.org/10.3389/fevo.2019.00091

Verant, Michelle L., et al. "White-nose syndrome initiates a cascade of physiologic disturbances in the hibernating bat host." *BMC Physiology*, 9 December 2014. doi.org/10.1186/s12899-014-0010-4

Voigt, Christian C., and Tigga Kingston. *Bats in the Anthropocene: Conservation of Bats in a Changing World*. Springer, 2015.

Windmill, James Frederick Charles, et al. "Keeping up with bats: dynamic auditory tuning in a moth." *Current Biology*, 19 December 2006. doi/10.1016/j.cub.2006.09.066

"Why are bats important?" U.S. Geological Survey. www.usgs.gov/faqs/why-are-bats-important

Foxfire Glowing

Anthony, Mark A., et al. "Enumerating soil biodiversity." *Proceedings of the National Academy of Sciences*, 7 August 2023. doi/abs/10.1073/pnas.2304663120

Balding, Mung, and Kathryn J. H. Williams. "Plant blindness and the implications for plant conservation." *Conservation Biology*, vol. 30, no. 6, 24 April 2016, pp. 1192–1199. doi/abs/10.1111/cobi.12738

"Blinded by the lights—nearly one in four drivers think most car headlights are too bright . . . and the problem is getting worse." *Drive*. Royal Automobile Club, 8 March 2022. rac.co.uk/drive/news/motoring-news/nearly-one-in-four-drivers-think-most-car-headlights-are-too-bright/

Brown, Clayton D. "North Carolina Rural Electrification: Precedent of the REA." *North Carolina Historical Review*, vol. 59, no. 2, April 1982, pp. 109–124. jstor.org/stable/23538636

Bush, Evan. "Amazon forests of the underground: Why scientists want to map the world's fungi." NBC News, 12 December 2021. nbcnews.com/science/science-news/amazon-forests-underground-scientists-want-map-worlds-fungi-rcna7899

Costanzi, Stefano. "Rhodopsin and the others: a historical perspective on structural studies of G protein-coupled receptors." *Current Pharmaceutical Design*, vol. 15, no. 35, pp. 3994–4002. doi/10.2174/138161209789824795

Cox, D. T. C., et al. "Majority of artificially lit Earth surface associated with the non-urban population." *Science of the Total Environment*, vol. 841, 1 October 2022. doi/10.1016/jscitotenv.2022.156782

Ekirch, Roger A. *At Day's Close: Night in Times Past*. Norton, 2005.

El Kouarti, Joyce. "The invasion of the forest destroyers—and how science is fighting back." U.S. Forest Service, 23 April 2021. fs.usda.gov/features/invasion-forest-destroyers-and-how-science-fighting-back

Gorzelak, Monika A. "Inter-plant communication through mycorrhizal networks mediates complex adaptive behaviour in plant communities." *AoB Plants*, 15 May 2015. doi/10.1093/aobpla/plv050

Hawksworth, David L., and Robert Lücking. "Fungal Diversity Revisited: 2.2 to 3.8 Million Species." *Microbiology Spectrum*, vol. 5, no. 4, 28 July 2017. doi/10.1128/microbiolspec.funk-0052-2016

Holmes, Rebecca. "Seeing single photons." *Physics World*, December 2016, pp. 28–31. research.physics.illinois.edu/QI/Photonics/pdf/PWDec16Holmes.pdf

Kraus, Louis J. "Human and Environmental Effects of Light Emitting Diode (LED) Community Lighting." American Medical Association Council on Science and Public Health, 2016. ama-assn.org/sites/ama-assn.org/files/corp/media-browser/public/about-ama/councils/Council%20Reports/council-on-science-public-health/a12-csaph4-lightpollution-summary.pdf

Kyba, Christopher C. M., et al. "Citizen scientists report global rapid reductions in the visibility of stars from 2011 to 2022." *Science*, 19 January 2023, vol. 379, no. 6629, pp. 265–268. doi/10.1126/science. abq7781

McCarthy, Erin. "Why Did Pirates Wear Eye Patches?" *Mental Floss*, 4 September 2013. mentalfloss.com/article/52493/why-did-pirates-wear-eye-patches

Muller, Joann, and Nathan Bomey. "Relief from ultra-bright headlights is coming—but slowly." *Axios*, 26 February 2023. axios.com/2023/02/26/car-headlights-too-bright-led

Parsley, Kathryn M. "Plant awareness disparity: A case for renaming plant blindness." *Plants People Planet*, vol. 2, no. 6, 3 October 2020, pp. 598–601. doi/10.1002/ppp3.10153

Purtov, Konstantin. "The Chemical Basis of Fungal Bioluminescence." *Angewandte Chemie International Edition*, 11 June 2015. doi/abs/10.1002/anie.201501779

"Rhodopsin." *ScienceDirect*. sciencedirect.com/topics/biochemistry-genetics-and-molecular-biology/rhodopsin

Ro, Christine. "Why 'plant blindness' matters—and what you can do about it." *BBC*, 24 February 2022. bbc.com/future/article/20190425-plant-blindness-what-we-lose-with-nature-deficit-disorder

"Terms of reference and rules of procedure of the World Forum for harmonization of vehicle regulations." United Nations, June 2019. un-ilibrary.org/content/books/9789210476478s010-c001

Wandersee, James H., and Elisabeth E. Schussler. "Preventing Plant Blindness." *American Biology Teacher*, vol. 61, no. 2, February 1999, pp. 82–86. doi/10.2307/4450624

Young, Robin, and Samantha Raphelson. "One Mycologist on Why Fungi Are Critical for the Survival of Life on This Planet." *Here & Now*. WBUR, 28 January 2019. wbur.org/hereandnow/2019/01/28/mushrooms-fungi-disease-bees

Moon Gardens Blooming

Ajmani, Gaurav S., et al. "Effects of Ambient Air Pollution Exposure on Olfaction: A Review." *Environmental Health Perspectives*, November 2016, pp. 1683–1693. doi/10.1289/EHP136

Bennie, Jonathan, et al. "Artificial light at night alters grassland vegetation species composition and phenology." *Journal of Applied Ecology*, 28 April 2017, pp. 442–450. doi/10.1111/1365-2664.12927

Berkowitz, Rachel. "Dark-sky advocates confront threats from above and below." *Physics Today*, 9 February 2022. doi.org/10.1063/PT.6.2.20220209b

Charlton, Anne. "Medicinal uses of tobacco in history." *Journal of the Royal Society of Medicine*, June 2004, pp. 292–296. doi/10.1258/jrsm.97.6.292

Clark, Stuart. "Are Elon Musk's 'megaconstellations' a blight on the night sky?" *Guardian*, 12 September 2020. theguardian.com/science/2020/sep/12/stars-astronomy-spacex-satellite-elon-musk

Czaja, Monika, and Anna Kolton. "How light pollution can affect spring development of urban trees and shrubs." *Urban Forestry & Urban Greening*, vol. 77, November 2022. doi/pii/S1618866722002965

Eichelberger, Julia. *Tell About Night Flowers: Eudora Welty's Gardening Letters 1940–1949*. University Press of Mississippi, 2015.

Falchi, Fabio. "A call for scientists to halt the spoiling of the night sky with artificial light and satellites." *Nature Astronomy*, vol. 7, pp. 237–239, March 2023. doi/10.1038/s41550-022-01864-z

ffrench-Constant, Richard H., et al. "Light Pollution is associated with earlier tree budburst across the United Kingdom." *Proceedings of the Royal Society*, vol. 283, no. 1833, 29 June 2016. doi/10.1098/rspb.2016.0813

Gaston, Kevin J. "The ecological impacts of nighttime light pollution: a mechanistic appraisal." *Biological Reviews*, 8 April 2013. doi/10.1111/brv.12036

Herz, Rachel S. "The Influence of Circadian Timing on Olfactory Sensitivity." *Chemical Senses*, vol. 43, no. 1, January 2018, pp. 45–51. doi/10.1093/chemse/bjx067

Hoover, Kara C., et al. "Global Survey of Variation in a Human Olfactory Receptor Gene Reveals Signatures of Non-neutral Evolution." *Chemical Senses*, vol. 40, no. 7, September 2015, pp. 481–488. doi/10.1093/chemse/bjv030

Jha, Shalene, et al. "Shade Coffee: Update on a Disappearing Refuge for Biodiversity." *BioScience*, vol. 64, no. 5, May 2024, pp. 416–428. doi/10.1093/biosci/biu038

Kahn Jr., Peter H., and Thea Weiss. "The Importance of Children Interacting with Big Nature." *Children, Youth, and Environments*, vol. 27, no. 2, 2017, pp. 7–24. jstor.org/stable/10.7721/chilyoutenvi.27.2.0007

Knop, Eva, et al. "Artificial light at night as a new threat to pollination." *Nature*, vol. 548, 10 August 2017, pp. 206–209. nature.com/articles/nature23288.epdf

Koutouleas, Athina, et al. "Shaded-Coffee: A Nature-based Strategy for Coffee Production Under Climate Change? A Review." *Frontiers Sustainable Food Systems*, 28 April 2022. doi/10.3389/fsufs.2022.877476/full

Loewer, Peter. *The Evening Garden*. Macmillan, 1993.

McGann, John P. "Poor human olfaction is a 19th-century myth." *Science*, vol. 356, 12 May 2017. doi/10.1126/science.aam7263

Meng, Lin, et al. "Artificial light at night: an underappreciated effect on phenology of deciduous woody plants." *Proceedings of the National Academy of Sciences Nexus*, vol. 1, no. 2, 18 April 2022. doi/10.1093/pnasnexus/pgac046

Nandakumar, Sangeetha, et al. "The high optical brightness of the BlueWalker3 satellite." *Nature*, vol. 623, 2 October 2023, pp. 938–941. doi.org/10.1038/s41586-023-06672-7

Shepherd, Tory. "Picture imperfect: light pollution from satellites is becoming an existential threat to astronomy." *Guardian*, 5 January 2023. theguardian.com/science/2023/jan/06/picture-imperfect-light-pollution-from-satellites-is-becoming-an-existential-threat-to-astronomy

Škvareninová, Jana, et al. "Effects of light pollution on tree phenology in the urban environment." *Moravian Geographical Reports*, vol. 25, no. 4, December 2017. doi.10.1515/mgr-2017-0024

"Traditional Tobacco." National Native Network. keepitsacred.itcmi.org/about-us/who-we-are/

Venkatesan, Aparna, et al. "The impact of satellite constellations on space as an ancestral global commons." *Nature Astronomy*, 6 November 2020, pp. 1043–1048. doi.org/10.1038/s41550-020-01238-3

von Arx, Martin, et al. "Floral humidity as a reliable sensory cue for profitability assessment by nectar-foraging hawkmoths." *Biological Sciences*, 29 May 2012. doi/abs/10.1073/pnas.1121624109

Wingrove, Jed, et al. "Aberrant olfactory network functional connectivity in people with olfactory dysfunction following COVID-19 infection: an exploratory, observational study." *eClinicalMedicine*, 2 March 2023. doi/10.1016/j.eclinm.2023.101883

Wolfe, Debbie. "Moon Gardens are the Dreamiest Plant Trend—Here's How to Create One." *Real Simple*, 9 July 2023. realsimple.com/home-organizing/gardening/outdoor/moon-garden

Zhang, Zhenyu. "Exposure to Particulate Matter Air Pollution and Anosmia." *JAMA Network*, May 2021. doi/10.1001/jamanetworkopen.2021.11606

Humans Surviving

Altermatt, Florian, and Dieter Ebert. "Reduced flight-to-light behaviour of moth populations exposed to long-term urban light pollution." *Biology Letters*, 1 April 2016, vol. 12, no. 4. doi/10.1098/rsbl.2016.0111

Barrett, Deirdre. *Supernormal Stimuli: How Primal Urges Overran Their Evolutionary Purpose*. W. W. Norton, 2010.

"Blue light has a dark side." *Staying Healthy*. Harvard Medical School, 7 July 2020. health.harvard.edu/staying-healthy/blue-light-has-a-dark-side

Brooks, Mike. "The Seductive Pull of Screens That You Might Not Know About." *Psychology Today*, 17 October 2018. psychologytoday.com/us/blog/tech-happy-life/201810/the-seductive-pull-screens-you-might-not-know-about

Cain, Sean W., et al. "Evening home lighting adversely impacts the circadian system and sleep." *Scientific Reports*, 5 November 2020. doi.org/10.1038/s41598-020-75622-4

Cherkasova, Mariya V., et al. "Win-concurrent Sensory Cues Can Promote Riskier Choice." *Journal of Neuroscience*, vol. 38, no. 48, 18 November 2018. doi.org/10.1523/JNEUROSCI.1171-18.2018

Fessler, Daniel. "A Burning Desire: Steps Toward an Evolutionary Psychology of Fire Learning." *Journal of Cognition and Culture*, September 2006. doi/10.1163/156853706778554986

Hartstein, Lauren. "High sensitivity of melatonin suppression response to evening light in preschool-aged children." *Journal of Pineal Research*, 7 March 2022. doi/10.1111/jpi.12780

Jones, Chas, et al. "Traditional Learnings: Into the Field with Yurok and U.S. Forest Service Experts on Cultural Burning of Forested Lands." *Partners, Science, People*. Northwest Climate Adaptation Science Center. nwcasc.uw.edu/2021/10/28/traditional-learnings-into-the-field-with-yurok-and-usfs-experts-on-cultural-burning-of-forested-lands/

Kaplan, Sarah. "How climbing down from trees let humans finally get a good night's sleep." *Washington Post*. 18 December 2015. Washingtonpost.com/news/morning-mix/wp/2015/12/18/how-climbing-down-from-trees-let-humans-finally-get-a-good-nights-sleep

Levy, Michelle. "How to Embrace the Benefits of Darkness." *Discover*, 19 August 2023. discovermagazine.com/planet-earth/how-to-embrace-the-benefits-of-darkness

London, Bianca. "Darkness Retreats Are Trending So What Are They?" *Glamour UK*, 16 May 2023. glamourmagazine.co.uk/article/darkness-retreats

Malůš, Marek, et al. "Darkness therapy as a tool for reducing anxiety, depression and for increasing well-being." STAR-Stress Anxiety Research Society Conference. July 2014. researchgate.net/publication/324918508_The_Darkness_therapy_as_a_tool_for_reducing_anxiety_depression_and_for_increasing_well-being_2014

Mason, Ivy C., et al. "Light exposure during sleep impairs cardiometabolic function." *Proceedings of the National Academy of Sciences*, 14 March 2022. doi/10.1073/pnas.2113290119

Moyes, Holley. *Sacred Darkness: A Global Perspective on the Ritual Use of Caves*. University Press of Colorado, 2012.

Moyes, Holley. "How caves showed me the connection between darkness and imagination." TEDx Talks, 22 November 2016. youtube.com/watch?v=TR13-JnS1ME

Muralidharan, Arumugam R., et al. "Light and myopia: from epidemiological studies to neurobiological mechanisms." *Therapeutic Advances in Ophthalmology*, December 2019. doi/10.1177/25158414211059246

Nelson, Randy J. "For Better Health, Think Paleo Lighting." *Huffington Post*, 11 October 2016. huffpost.com/entry/for-better-health-think-paleo-lighting_b_57fce3f6e4b0e655eab77b44

"New study reveals that exposure to outdoor artificial light at night is associated with an increased risk of diabetes." *Science Daily*, 14 November 2022. sciencedaily.com/releases/2022/11/221114190650.htm

Obayashi, Kenji, et al. "Bedroom Light Exposure at Night and the Incidence of Depressive Symptoms: A Longitudinal Study of the HEIJO-KYO Cohort." *American Journal of Epidemiology*, 1 March 2018. doi/10.1093/aje/kwx290

Papadopoulos, Costas, and Holley Moyes, editors. *The Oxford Handbook of Light in Archaeology*, 6 June 2017. Oxford University Press.

Park, Yong-Moon Mark, et al. "Association of Exposure to Artificial Light at Night While Sleeping with Risk of Obesity in Women." *JAMA Internal Medicine*, 1 August 2019. Doi/10.1001/jamainternmed.2019.0571

Peintner, U., et al. "The iceman's fungi." *Mycological Research*, vol. 102, no. 10, October 1998, pp. 1153–1162. doi/10.1017/S0953756298006546

Pyne, Stephen. *Fire: A Brief History*. University of Washington Press, 2019.

Siraji, Mushfigul Anwar, et al. "Light exposure behaviors predict mood, memory, and sleep quality." *Scientific Reports*, 1 August 2023. doi.org/10.1038/s41598-023-39636-y

Spence, Charles, and Betina Piqueras-Fiszman. "Dining in the Dark." *Psychologist*, 16 December 2012. bps.org.uk/psychologist/dining-dark

Wang, Robin R. *Yinyang: The Way of Heaven and Earth in Chinese Thought and Culture*. Cambridge University Press, 2012.

Wang, Robin R. "Yinyang." *Encyclopedia of Philosophy.* IEP. iep.utm.edu/yinyang/

Watts, Steven M. *Practicing Primitive: A Handbook of Aboriginal Skills.* Gibbs Smith, 2005.

Wiessner, Polly W. "Embers of society: Firelight talk among the Ju/'hoansi Bushmen." *Proceedings of the National Academy of Sciences*, 22 September 2014, pp. 14027–14035. doi/full/10.1073/pnas.1404212111

Additional Resources

American Bird Conservancy | abcbirds.org

Arc of Appalachia | Arcofappalachia.org

Bat Conservation International | Batcon.org

BirdCast | Birdcast.info

Blue Marble Navigator | blue-marble.de/nightlights/

Cornell Lab of Ornithology | Birds.cornell.edu

DarkSky International | Darksky.org

Discover Life in America | Dlia.org

Firefly Conservation & Research | Firefly.org

Fireflyers International Network | fireflyersinternational.net

Fire Learning Network | firenetworks.org

Foxfire Fund | Foxfire.org

Fungi Foundation | Ffungi.org

Globe at Night | Globeatnight.org

Grandfather Mountain Foundation | Grandfather.com/foundation

Homegrown National Park | Homegrownnationalpark.org

iNaturalist | inaturalist.org

National Audubon Society | audubon.org

National Moth Week | Nationalmothweek.org

North American Native Plant Society | Nanps.org

Primitive Skill Gatherings | paleotechnics.com/gatherings

Satellite Streak Watcher | anecdata.org/projects/view/687

Society for the Protection of Underground Networks | spun.earth

University of Utah, Dark Sky Studies | advising.utah.edu/majors/quick-look/dark-skies.php

Vernal Pool Association | Vernalpool.org

Xerces Society for Invertebrate Conservation | Xerces.org